Undergraduate

Editors
S. Axler
F. W. Gehring
K. A. Ribet

Springer
New York
Berlin
Heidelberg
Hong Kong
London
Milan
Paris
Tokyo

Undergraduate Texts in Mathematics

(continued after index)

M. A. Armstrong

Basic Topology

With 132 Illustrations

 Springer

M. A. Armstrong
Department of Mathematics
University of Durham
Durham DH1 3LE
England

Mathematics Subject Classification (2000): 55-01, 54-01

Library of Congress Cataloging-in-Publication Data
Armstrong, M. A. (Mark Anthony)
    Basic topology.
    (Undergraduate texts in mathematics)
    Bibliography: p.
    Includes index.
    1. Topology.   I. Title.   II. Series.
QA611.A68   1983   514   83-655

Original edition published in 1979 by McGraw-Hill Book Company (UK) Ltd., Maidenhead,
Berkshire, England.

ISBN 978-1-4419-2819-1

Printed in the United States of America.

9 8

Springer-Verlag is a part of *Springer Science+Business Media*

springeronline.com

Dedicated to the memory of

PAUL DUFÉTELLE

Preface to the Springer Edition

This printing is unchanged, though the opportunity has been taken to correct one or two misdemeanours. In particular Problems 2.13, 3.13 and 3.19 are now correctly stated, and Tietze has regained his final "e". My thanks go to Professor P. R. Halmos and to Springer-Verlag for the privilege of appearing in this series.

M.A.A.
Durham, January 1983.

Preface

This is a topology book for undergraduates, and in writing it I have had two aims in mind. Firstly, to make sure the student sees a variety of different techniques and applications involving point set, geometric, and algebraic topology, without delving too deeply into any particular area. Secondly, to develop the reader's geometrical insight; topology is after all a branch of geometry.

The prerequisites for reading the book are few, a sound first course in real analysis (as usual!), together with a knowledge of elementary group theory and linear algebra. A reasonable degree of 'mathematical maturity' is much more important than any previous knowledge of topology.

The layout is as follows. There are ten chapters, the first of which is a short essay intended as motivation. Each of the other chapters is devoted to a single important topic, so that identification spaces, the fundamental group, the idea of a triangulation, surfaces, simplicial homology, knots and covering spaces, all have a chapter to themselves.

Some motivation is surely necessary. A topology book at this level which begins with a set of axioms for a topological space, as if these were an integral part of nature, is in my opinion doomed to failure. On the other hand, topology should not be presented as a collection of party tricks (colouring knots and maps, joining houses to public utilities, or watching a fly escape from a Klein bottle). These things all have their place, but they must be shown to fit into a unified mathematical theory, and not remain dead ends in themselves. For this reason, knots appear at the end of the book, and not at the beginning. It is not the knots which are so interesting, but rather the variety of techniques needed to deal with them.

Chapter 1 begins with Euler's theorem for polyhedra, and the theme of the book is the search for topological invariants of spaces, together with techniques for calculating them. Topology is complicated by the fact that something which is, by its very nature, topologically invariant is usually hard to calculate, and vice versa the invariance of a simple number like the Euler characteristic can involve a great deal of work.

The balance of material was influenced by the maxim that a theory and its payoff in terms of applications should, wherever possible, be given equal weight. For example, since homology theory is a good deal of trouble to set up (a whole chapter), it must be shown to be worth the effort (a whole chapter of applications). Moving away from a topic is always difficult, and the temptation to include more and more is hard to resist. But to produce a book of reasonable length some topics just have to go; I mention particularly in this respect the omission of any systematic method for calculating homology groups. In formulating definitions, and choosing proofs, I have not always taken the shortest path. Very often the version of a definition or result which is most convenient to work with, is not at all natural at first sight, and this is above all else a book for beginners.

Most of the material can be covered in a one-year course at third-year (English) undergraduate level. But there is plenty of scope for shorter courses involving a selection of topics, and much of the first half of the book can be taught to second-year students. Problems are included at the end of just about every section, and a short bibliography is provided with suggestions for parallel reading and as to where to go next.

The material presented here is all basic and has for the most part appeared elsewhere. If I have made any contribution it is one of selection and presentation.

Two topics deserve special mention. I first learned about the Alexander polynomial from J. F. P. Hudson, and it was E. C. Zeeman who showed me how to do surgery on surfaces. To both of them, and particularly to Christopher Zeeman for his patience in teaching me topology, I offer my best thanks.

I would also like to thank R. S. Roberts and L. M. Woodward for many useful conversations, Mrs J. Gibson for her speed and skill in producing the manuscript, and Cambridge University Press for permission to reproduce the quotation from Hardy's 'A Mathematician's Apology' which appears at the beginning of Chapter 1. Finally, a special word of thanks to my wife Anne Marie for her constant encouragement.

M.A.A.
Durham, July 1978.

Contents

1. Introduction

*Beauty is the first test: there is no permanent place in the
world for ugly mathematics.*

G. H. HARDY

1.1 Euler's theorem

We begin by proving a beautiful theorem of Euler concerning polyhedra. As we shall see, the statement and proof of the theorem motivate many of the ideas of topology.

Figure 1.1 shows four polyhedra. They look very different from one another,

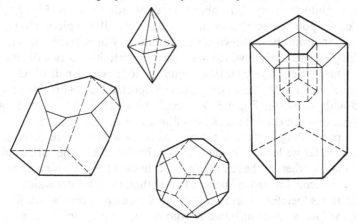

Figure 1.1

yet if for each one we take the number of vertices (v), subtract from this the number of edges (e), then add on the number of faces (f), this simple calculation always gives 2. Could the formula $v - e + f = 2$ be valid for all polyhedra? The answer is no, but the result is true for a large and interesting class.

We may be tempted at first to work only with regular, or maybe convex, polyhedra, and $v - e + f$ is indeed equal to 2 for these. However, one of the examples in our illustration is not convex, yet it satisfies our formula and we would be unhappy to have to ignore it. In order to find a counterexample we need to be a little more ingenious. If we do our calculation for the polyhedra shown in Figs 1.2 and 1.3 we obtain $v - e + f = 4$ and $v - e + f = 0$ respect-

1

ively. What has gone wrong? In the first case we seem to have cheated a little by constructing a polyhedron whose surface consists of two distinct pieces; in

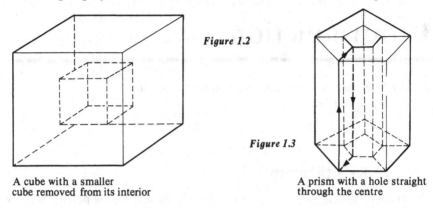

Figure 1.2

Figure 1.3

A cube with a smaller
cube removed from its interior

A prism with a hole straight
through the centre

technical language its surface is not connected. We suspect (quite correctly) that we should not allow this, since each of the pieces of surface contributes 2 to $v - e + f$. Unfortunately, this objection does not hold for Fig. 1.3, as the surface of the polyhedron shown there is certainly all one piece. However, this surface differs from those shown earlier in one very important respect. We can find a loop on the surface which does not separate it into two distinct parts; that is to say, if we imagine cutting round the loop with a pair of scissors then the surface does not fall into two pieces. A specific loop with this property is labelled with arrows in Fig. 1.3. We shall show that $v - e + f = 2$ for polyhedra which do not exhibit the defects illustrated in Figs 1.2 and 1.3.

Before proceeding any further, we need to be a little more precise. In our discussion so far we have only made use of the surfaces of the solids illustrated (except, that is, when we have mentioned convexity). So let us agree to use the word 'polyhedron' for such a surface, rather than for the solid which it bounds. A *polyhedron* is therefore a finite collection of plane polygons which fit together nicely in the following sense. If two polygons meet they do so in a common edge, and each edge of a polygon lies in precisely one other polygon. In addition, we ask that if we consider the polygons which contain a particular vertex, then we can label them $Q_1, Q_2, \ldots, Q_k$ in such a way that Q_i has an edge in common with Q_{i+1} for $1 \leqslant i < k$, and Q_k has an edge in common with Q_1. In other words, the polygons fit together to form a piece of surface around the given vertex. (The number k may vary from one vertex to another.) This last condition rules out, for example, two cubes joined together at a single vertex.

(1.1) Euler's theorem. *Let* P *be a polyhedron which satisfies:*
(a) *Any two vertices of* P *can be connected by a chain of edges.*
(b) *Any loop on* P *which is made up of straight line segments (not necessarily edges) separates* P *into two pieces.*
Then v − e + f = 2 *for* P.

2

The formula $v - e + f = 2$ has a long and complicated history. It first appears in a letter from Euler to Goldbach dated 1750. However, Euler placed no restrictions on his polyhedra and his reasoning can only be applied in the convex case. It took sixty years before Lhuilier drew attention (in 1813) to the problems raised by polyhedra such as those shown in our Figs 1.2 and 1.3. The precise statement of theorem (1.1), and the proof outlined below, are due to von Staudt and were published in 1847.

Outline proof. A connected set of vertices and edges of P will be called a *graph*: connected simply means that any two vertices can be joined by a chain of edges in the graph. More generally, we shall use the word graph for any finite connected set of line segments in 3-space which fit together nicely as in Fig. 1.4. (If two segments intersect they are required to do so in a common vertex.) A graph which does not contain any loops is called a *tree*. Notice that for a tree, the number of vertices minus the number of edges is equal to 1. If the tree is denoted by T we shall write this as $v(T) - e(T) = 1$.

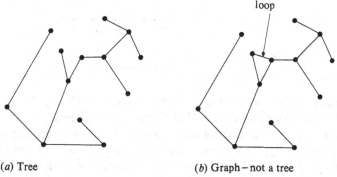

(a) Tree   (b) Graph − not a tree

Figure 1.4

By hypothesis (a), the set of all vertices and edges of P is a graph. It is easy to show that in any graph one can find a subgraph which is a tree and which contains all the vertices of the original. So choose a tree T which consists of some of the edges and *all* of the vertices of P (Fig. 1.4a shows such a tree for one of the polyhedra of Fig. 1.1).

Now form a sort of 'dual' to T. This dual is a graph Γ defined as follows. For each face A of P we give Γ a vertex $\hat{A}$. Two vertices $\hat{A}$ and $\hat{B}$ of Γ are joined by an edge if and only if the corresponding faces A and B of P are adjacent with intersection an edge that is not in T. One can even represent Γ on P in such a way that it misses T (the vertex $\hat{A}$ corresponding to an interior point of A) though to do this we have to allow its edges to be bent. Figure 1.5 illustrates the procedure.

It is not too hard to believe that this dual Γ is connected and is therefore a graph. Intuitively, if two vertices of Γ cannot be connected by a chain of edges of Γ, then they must be separated from one another by a loop of T. (This does

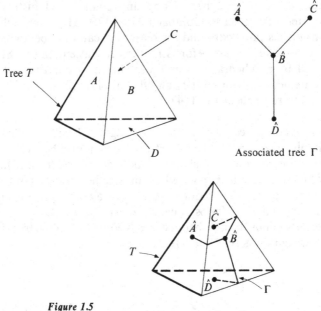

Figure 1.5

Tree T

Associated tree Γ

Γ represented on P

need some proof and we shall work out the details in Chapter 7.) Since T does not contain any loops we deduce that Γ must be connected.

In fact Γ is a tree. For if there were a loop in Γ it would separate P into two distinct pieces by hypothesis (b), and each of these pieces must contain at least one vertex of T. Any attempt to connect two vertices of T which lie in different pieces by a chain of edges results in a chain which meets this separating loop, and therefore in a chain which cannot lie entirely in T. This contradicts the fact that T is connected. Therefore Γ is a tree. (The proof breaks down here for a polyhedron such as that shown in Fig. 1.3, because the dual graph Γ will contain loops.)

Since the number of vertices of any tree exceeds the number of edges by 1 we have $v(T) - e(T) = 1$ and $v(\Gamma) - e(\Gamma) = 1$. Therefore

$$v(T) - [e(T) + e(\Gamma)] + v(\Gamma) = 2.$$

But by construction $v(T) = v$, $e(T) + e(\Gamma) = e$ and $v(\Gamma) = f$. This completes the argument.

1.2 Topological equivalence

There are several proofs of Euler's theorem. We have chosen the one above for two reasons. Firstly, its elegance; most other proofs use induction on the number of faces of P. Secondly, because it contains much more information than Euler's formula. With very little extra effort it actually tells us that P is made up of two discs which are identified along their boundaries. To see this, simply thicken each of T and Γ a little on P (Fig. 1.6) to obtain two disjoint discs. (Thickening a tree always gives a disc, though thickening a graph with loops will give a space with holes in it.) Enlarge these discs little by little until their boundaries coincide. The polyhedron P is now made up of two discs which have a common boundary. Granted these discs may have a rather odd shape, but they can be deformed into ordinary, round flat discs. Now remember that the sphere consists of two discs, the north and south hemispheres, sewn along their common boundary the equator (Fig. 1.7). In other words, the hypotheses of Euler's theorem tell us that P looks in some sense like a rather deformed sphere.

T and Γ thickened on P

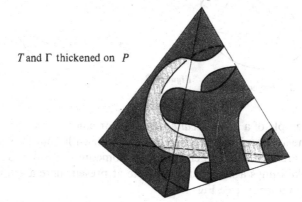

Figure 1.6

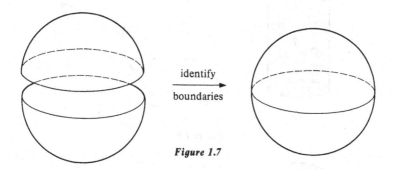

identify

⟶

boundaries

Figure 1.7

Of course, for a specific polyhedron it may be very easy to set up a decent correspondence between its points and those of the sphere. For example, in the case of the regular tetrahedron T we can use radial projection from the centre of gravity $\hat{T}$ of T to project T onto a sphere with centre $\hat{T}$. The faces of T project to curvilinear triangles on the sphere as shown in Fig. 1.8. In fact Legendre used exactly this procedure (in 1794) to prove Euler's theorem for convex polyhedra; we shall describe Legendre's argument later.

The polyhedron shown on the right in Fig. 1.1 is not convex and does not lend itself to the above argument. However, if we think of it as being made of rubber then we can easily imagine how to deform it into an ordinary round sphere. During the deformation we stretch and bend the polyhedron at will, *but we never identify distinct points and we never tear it.* The resulting correspondence between the points of the given polyhedron and the points of the

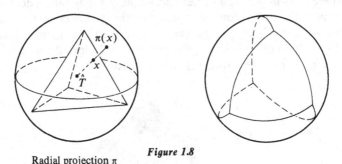

Radial projection π

Figure 1.8

sphere is an example of a *topological equivalence* or *homeomorphism*. In formal terms it is a one–one and onto continuous function with continuous inverse.

We shall go carefully into the definition of a homeomorphism in Section 1.4, but to help make things a little more concrete at present here are four spaces which are homeomorphic (see Fig. 1.9):

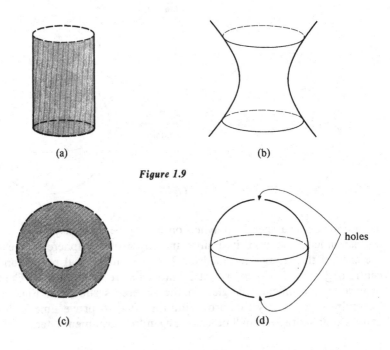

(a)

(b)

Figure 1.9

(c)

(d)

holes

(a) the surface of a cylinder of finite height, excluding the two circles at the ends;
(b) the one-sheeted hyperboloid given by the equation $x^2 + y^2 - z^2 = 1$;
(c) the open annulus in the complex plane specified by $1 < |z| < 3$;
(d) the sphere with the points at the north and south poles removed.

We propose to give a specific homeomorphism (i.e., a continuous, one–one, and onto function which has continuous inverse) from space (b) to space (c). It is most convenient to specify the points of (b) by cylindrical polar coordinates (r, θ, z) and to use plane polar coordinates (r, θ) for space (c). When $\theta = 0$ in (b) we obtain a branch of the hyperbola $x^2 - z^2 = 1$, and we plan to send this nicely onto the corresponding piece of the annulus, i.e., the ray $\{(x,y) \mid 1 < x < 3, y = 0\}$. If we can do a similar trick for each value of θ, in a continuous manner as θ varies from 0 to 2π, we shall have the required homeomorphism. Define $f:(-\infty, \infty) \to (1, 3)$ by $f(x) = x/(1 + |x|) + 2$; then f is a bijection, is continuous, and has continuous inverse. Now send the point (r, θ, z) of the hyperbola to $(f(z), \theta)$ on the annulus.

We leave the reader to investigate the other possibilities: we note that the relation of topological equivalence is clearly an equivalence relation, so that proving each of spaces (a) and (d) homeomorphic to space (c) will suffice. In topology these four are considered to be the 'same space'. The sphere with three points removed is different (not homeomorphic to the above). Why? Can you describe a subset of the complex plane homeomorphic to a sphere with three points removed?

Returning to the proof of Euler's theorem, thickening the trees T and Γ gave a decomposition of P into two discs with a common boundary and therefore, by sending the points of one disc into the northern hemisphere and sending the points of the other south, a way of defining a homeomorphism from the polyhedron P to the sphere. It is possible to produce an argument in the opposite direction (we shall do so in Chapter 7) and show that if P is topologically equivalent to the sphere then P satisfies hypotheses (a) and (b) of theorem (1.1)†, and therefore Euler's theorem holds for P. So if P and Q are polyhedra which are both homeomorphic to the sphere, and if we call $v - e + f$ the *Euler number* of a polyhedron, then we know from the above discussion that P and Q have the same Euler number, namely 2.

The polyhedron shown in Fig. 1.3 has an entirely different form. It is homeomorphic to a torus (we can even imagine how to deform it continuously to a nice round torus such as that shown in Fig. 1.10b) and its Euler number is 0. Drawing any other polyhedron which is topologically equivalent to a torus and computing its Euler number will always give 0 (though this is hard to prove and will have to wait until Chapter 9). We are now only a short step‡ away from one of the most basic and central results of topology.

---

† Hypothesis (a) is easy to verify; (b) is harder and is a special case of the famous Jordan curve theorem.

‡ A short step, that is, in mathematical intuition; in terms of careful proof we have an extremely long walk ahead.

(1.2) Theorem. *Topologically equivalent polyhedra have the same Euler number.*

This remarkable result was the starting point for modern topology. It is remarkable because in calculating the Euler number of a polyhedron we make use of the numbers of vertices, edges, and faces of the polyhedron, none of which need be preserved by a topological equivalence. It led to the search for other properties of spaces which are left unchanged by the application of a homeomorphism.

We shall return to the Euler number later, and show that it can be defined for a much wider class of spaces than the ordinary polyhedra considered so far. These polyhedra are rather rigid objects with corners, edges, and flat faces, and will be of no special interest to us. From the point of view of the topologist, the sphere is good enough to represent all the polyhedra shown in Fig. 1.1. Our philosophy will be roughly as follows: the Euler number 2 does not belong to a particular class of polyhedra, *it really belongs to the sphere.* A polyhedron satisfying the hypotheses of Euler's theorem (i.e., a polyhedron which is homeomorphic to the sphere) merely gives a convenient way of calculating the Euler number of the sphere. With this emphasis, theorem (1.2) now states that calculating in apparently different ways always gives the same answer. We shall continue this line of argument in Section 9.2.

We end this section with Legendre's highly original proof of Euler's formula for convex polyhedra. Using radial projection as in Fig. 1.8, project the polyhedron onto a sphere of radius 1. The polygonal faces of the polyhedron project to spherical polygons. Now if Q is a spherical polygon with angles $\alpha_1, \alpha_2, \ldots, \alpha_k$ and with n edges, then the area of Q is given by

$$\alpha_1 + \alpha_2 + \ldots + \alpha_k - (n-2)\pi = (\alpha_1 + \alpha_2 + \ldots + \alpha_k) - n\pi + 2\pi$$

The sum of the areas of the spherical polygons is therefore $2\pi v - 2\pi e + 2\pi f$ (at each vertex the total angle is 2π, so $2\pi v$ takes care of all the α's; each edge has to be counted twice since it belongs to exactly two polygons; and each face gives a contribution of 2π). Equating this to the area of the unit sphere, 4π, gives the result.

1.3 Surfaces

Topology has to do with those properties of a space which are left unchanged by the kind of transformation that we have called a topological equivalence or homeomorphism. But what sort of spaces interest us and what exactly do we mean by a 'space'? The idea of a homeomorphism involves very strongly the notion of continuity; what do we mean by a continuous function between two spaces? We shall try to answer these questions in this section and the next.

We begin with a few examples of interesting spaces. Someone working in analysis is used to considering the real line, the complex plane, or even the set of all real-valued continuous functions defined on the closed unit interval as (metric) spaces. Being geometers at heart, we are more interested in bounded configurations which occur naturally in euclidean space. For example, the unit circle and the unit disc in the plane; surfaces such as the sphere, the torus, the Möbius strip, the cylinder, and the double torus with a puncture, all of which live in three-dimensional space and are illustrated in Fig. 1.10.

Of a more complicated nature and more difficult to visualize is a surface such as the Klein bottle. This surface is difficult to imagine because in any attempt to represent it in three dimensions the Klein bottle must cross itself. In our drawing (Fig. 1.10) the surface cuts itself in a small circle. We can understand the Klein bottle a little better by trying to make a model of it. Consider the usual method of modelling a torus. One begins with a rectangle of paper and

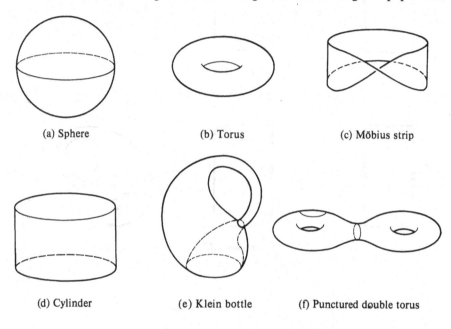

(a) Sphere	(b) Torus	(c) Möbius strip

(d) Cylinder	(e) Klein bottle	(f) Punctured double torus

Figure 1.10

identifies the edges as in Fig. 1.11. To build a Klein bottle, the first half of the construction, that is as far as a cylinder, is the same, but then the ends of the cylinder are identified in the opposite direction. In order to do this, the cylinder has to be bent around and one end pushed through the side as in Fig. 1.12.

The Klein bottle (K) can be represented in four-dimensional space without any self intersections. Imagine an extra dimension perpendicular to the paper, remembering all the time that the paper represents ordinary three-dimensional space. Near the intersection circle of K we have two pipes, one of which cuts through the other. Lift one pipe a little clear of the other into the fourth dimension. If you find this hard, examine the following procedure which is easier to visualize: Fig. 1.13a shows two lines in the plane which cross at right angles. Suppose we wish to move them *very slightly* so that they no longer intersect.

It is obvious that we cannot do this in the plane. However, adding an extra dimension perpendicular to the plane of the paper gives us three-dimensional space, and we can simply lift one line a very small amount near the intersection point in this extra direction. This gives the two lines of Fig. 1.13b which no longer meet.

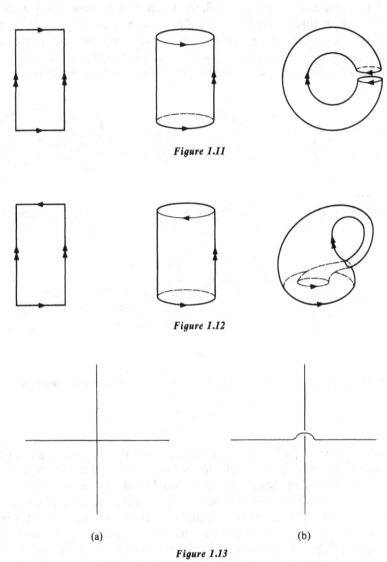

Figure 1.11

Figure 1.12

(a)   (b)

Figure 1.13

Our way of introducing surfaces by representing them in euclidean space is not such a good idea as it might appear at first sight. We are interested in surfaces 'up to alteration by a homeomorphism', in other words topologically equivalent surfaces will be treated as the *same* space. We illustrate in Fig. 1.14 three copies of the Möbius strip *M*. That the first two are homeomorphic is no surprise, one only has to take a rubber version† of the first and stretch it into the second. But how about Figs 1.14a and 1.14c? These spaces *are* homeomorphic, yet no amount of stretching, bending, and twisting will deform one into the other. To show two spaces are homeomorphic one must find a continuous bijection between them, whose inverse is also continuous. Forget about

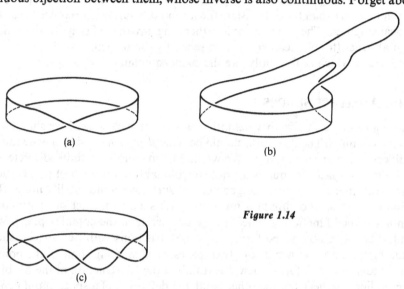

(a)

(b)

Figure 1.14

(c)

the various pictures of *M* and ask yourself how you construct *M*. Building a model is easy: begin with a rectangle of paper and identify a pair of opposite edges with a half twist (Fig. 1.15). This gives the usual representation of *M*

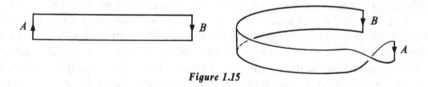

Figure 1.15

shown in Fig. 1.14a. To obtain Fig. 1.14c we must add a full twist to the above process, i.e., identify the edges of our strip after twisting one and one-half times. But in terms of the identification of the edges *A* and *B* this *changes nothing*, the same points of *A* and *B* are made to coincide. Therefore the spaces shown in Figs 1.14a and 1.14c are homeomorphic. They are merely different representations of the *same* space in euclidean space. The representations are different in

---

† The idea of explaining topological equivalence by thinking of spaces as being made of rubber is due to Möbius and dates back to about 1860.

the sense that although we can find a homeomorphism between them, there is no way of extending such a homeomorphism over all the points of euclidean space. In other words, there is no homeomorphism from euclidean space to itself that throws Fig. 1.14a onto Fig. 1.14c.

If our intuition can be led astray by pictures as naive as Fig. 1.14, this suggests very strongly that we need some way of considering our spaces *abstractly* rather than relying on particular representatives of them in euclidean space. In what follows we shall try to translate the notion of a surface into precise mathematical language. The programme is rather long, involving firstly the definition of an abstract (topological) space, and secondly the recognition of surfaces as those spaces which look locally like the euclidean plane.

1.4 Abstract spaces

In trying to find a satisfactory definition of a topological space† we shall have two aims in mind. The definition should be *general enough* to allow a wide range of different structures as spaces. We would like to consider a finite, discrete set of points as a space, or equally a whole uncountable continuum of points such as the real line; one of our nice geometrical surfaces should qualify under the definition, and so too should a function space such as the set of continuous complex-valued functions defined on the unit circle in the complex plane. We would like to be able to perform simple constructions with our spaces, such as taking the cartesian product of two spaces, or identifying some of the points of a space in order to form a new one (think of the construction of the Möbius strip outlined earlier). On the other hand, the definition of a space should contain *enough information* so that we can define the notion of continuity for functions between spaces. It is really this second consideration which leads to the abstract definition given below.

Let f be a function between two euclidean spaces, say $f: \mathbb{E}^m \to \mathbb{E}^n$. The classical definition of continuity for f goes as follows: f is continuous at $\mathbf{x} \in \mathbb{E}^m$ if given $\varepsilon > 0$ there exists $\delta > 0$ such that $\| f(\mathbf{y}) - f(\mathbf{x}) \| < \varepsilon$ whenever $\| \mathbf{y} - \mathbf{x} \| < \delta$. The function f is continuous if it satisfies this condition for each $\mathbf{x}$ in $\mathbb{E}^m$. Call a subset of N of $\mathbb{E}^m$ a *neighbourhood* of the point $\mathbf{p} \in \mathbb{E}^m$ if for some real number $r > 0$ the closed disc centre $\mathbf{p}$ radius r lies entirely inside N. It is easy to rephrase the above definition of continuity as follows: f is continuous if given any $\mathbf{x} \in \mathbb{E}^m$ and any neighbourhood N of $f(\mathbf{x})$ in $\mathbb{E}^n$, then $f^{-1}(N)$ is a neighbourhood of $\mathbf{x}$ in $\mathbb{E}^m$.

This notion of each point in a space having a collection of 'neighbourhoods', the neighbourhoods leading in turn to a good definition of continuous function, is the crucial one. Notice that in defining neighbourhoods in a euclidean space we used very strongly the euclidean distance between points. In constructing an abstract space we would like to retain the concept of neighbourhood but *rid*

---

† The modern definition emerged quite late, the axioms for a topological space appearing for the first time in 1914 in the work of Hausdorff.

ourselves of any dependence on a distance function. (A topological equivalence does not preserve distances.)

Inspection of the properties of neighbourhoods of points in a euclidean space leads to the following axioms for a topological space.

(1.3) We ask for a set X and for each point x of X a nonempty collection of subsets of X, called neighbourhoods of x. These neighbourhoods are required to satisfy four axioms:

(a) x *lies in each of its neighbourhoods.*
(b) *The intersection of two neighbourhoods of* x *is itself a neighbourhood of* x.
(c) *If* N *is a neighbourhood of* x *and if* U *is a subset of* X *which contains* N, *then* U *is a neighbourhood of* x.
(d) *If* N *is a neighbourhood of* x *and if* $\mathring{N}$ *denotes the set* $\{z \in N \,|\, N$ *is a neighbour-hood of* z$\}$, *then* $\mathring{N}$ *is a neighbourhood of* x. (The set $\mathring{N}$ is called the *interior* of N.)

This whole structure is called a *topological space*. The assignment of a collection of neighbourhoods satisfying axioms (a)–(d) to each point $x \in X$ is called a *topology* on the set X. (To provide a little motivation for axiom (d) take a point x in $\mathbb{E}^m$ and let B (for ball) denote the set of points distance less than or equal to 1 from x. Then B is a neighbourhood of x. The interior of B is simply those points distance less than 1 from x (the ball minus its boundary) and is still a neighbourhood of x.)

We can now say precisely what we mean by a continuous function and by a homeomorphism. Let X and Y be topological spaces†. A function $f : X \to Y$ is *continuous* if for each point x of X and each neighbourhood N of $f(x)$ in Y the set $f^{-1}(N)$ is a neighbourhood of x in X. A function $h : X \to Y$ is called a homeomorphism if it is one–one, onto, continuous, and has a continuous inverse. When such a function exists, X and Y are called *homeomorphic* (or *topologically equivalent*) spaces.

Suddenly things have become very complicated; we need a few examples in order to clear the air and help our intuition along.

Examples.

1. Any euclidean space with the usual definition of neighbourhood is a topological space. We shall show later that *euclidean spaces of different dimensions cannot be homeomorphic.* This is a hard problem, but its solution is essential if we are to have any confidence that our definition of homeomorphism can survive happily alongside our idea of the dimension of a space.

2. Let X be a topological space and let Y be a subset of X. We can define a topology on Y as follows. Given a point $y \in Y$ take the collection of its neighbourhoods in the topological space X and intersect each of these neighbour-

---

† So each of the letters X and Y represents a lot of information, namely the complicated structure of definition (1.3).

hoods with Y. The resulting sets are the neighbourhoods of y in Y. The axioms for a topology are easily checked and we say that Y has the *subspace topology*. This is a very useful procedure; it allows us, for example, to consider any subset of a euclidean space as a topological space. *In particular, our examples of surfaces become topological spaces.*

3. Let C denote the unit circle in the complex plane and $[0,1)$ those real numbers which are greater than or equal to 0 and less than 1. Give both of these sets the subspace topology from the plane and the real line respectively. Defining $f:[0,1) \to C$ by $f(x) = e^{2\pi i x}$ gives an example of a continuous function. Note that this function is one–one and onto. Its inverse is *not* continuous. (Why not?) This illustrates very well the importance of the condition that the inverse function be continuous in the definition of a homeomorphism: we would after all be very unhappy if the circle turned out to be homeomorphic to an interval.

4. Take the situation shown in Fig. 1.8 and consider the sphere and the surface of the tetrahedron as subspaces of $\mathbb{E}^3$. Check that radial projection π gives a homeomorphism between these two spaces. This type of homeomorphism is called a *triangulation* (of the sphere in this case) and will be the subject of a later chapter.

5. A distance function or metric on a set gives rise to a topology on the set. The construction of the neighbourhoods is entirely analogous to the procedure in a euclidean space. We illustrate the situation for a space of functions. Let X be the set of all continuous real-valued functions defined on a closed interval I of the real line. A function in the set is necessarily bounded and the usual distance function on X is defined by

$$d(f,g) = \sup_{x \in I} |f(x) - g(x)|.$$

Given a function $f \in X$, a subset N of X is a neighbourhood of f if for some positive real number ε the collection of all functions distance less than or equal to ε from f lies inside N.

6. Two different topological spaces may have the same underlying set of points. As an example of a rather peculiar topology on the set of real numbers, define a subset of the reals to be a neighbourhood of a particular real number if it contains that number and if in addition its complement is finite. This gives a topological space very different from (not homeomorphic to) the real line. Notice that no distance function on the set of real numbers can give rise to this topology. (Why not?)

7. Let X be a set and for each point $x \in X$ define $\{x\}$ to be a neighbourhood of x. So by axiom (c), any subset of X which contains x is a neighbourhood of x. Intuitively we think of this topology as making X into a discrete set of points – we have arranged for each point x to have a neighbourhood that contains *no other points*. With this topology *any* function with domain X is continuous.

We have now developed enough machinery to say exactly what we mean by a surface, and free ourselves from the straightjacket of having to work inside some euclidean space.

(1.4) Definition. *A surface is a topological space in which each point has a neighbourhood homeomorphic to the plane, and for which any two distinct points possess disjoint neighbourhoods.*

It is worth taking the time to examine this definition in some detail. The requirement that each point of the space should have a neighbourhood which is homeomorphic to the plane fits exactly our intuitive idea of what a surface should be. If we stand in it at some point (imagining a giant version of the surface in question) and look at the points very close to our feet we should be able to imagine that we are standing on a plane. The surface of the earth is a good example. Unless you belong to the Flat Earth Society you believe it to be (topologically) a sphere, yet locally it looks distinctly planar. Think more carefully about this requirement: we ask that some neighbourhood of each point of our space be homeomorphic to the plane. We have then to treat this neighbourhood as a topological space in its own right. But this presents no difficulty; the neighbourhood is after all a subset of the given space and we can therefore supply it with the subspace topology.

The second requirement, that any two distinct points possess disjoint neighbourhoods, is more technical in nature. It is motivated by our experience: all of our examples of surfaces have this property; unfortunately it is not automatically satisfied by spaces which locally look like the plane.

We have given the simplest possible definition. If we wish to allow a surface to have an edge or boundary (as in the case of the Möbius strip), then we cannot expect every point to have a neighbourhood homeomorphic to the plane. We must allow in addition points which have neighbourhoods homeomorphic to

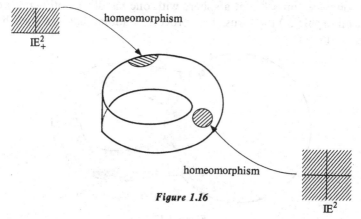

Figure 1.16

the upper half-plane (consisting of those points of the plane whose y-coordinates are greater than or equal to zero). All of our examples of surfaces now fit in nicely with this definition when they are given the subspace topology from euclidean space. Figure 1.16 illustrates the definition for a Möbius strip.

1.5 A classification theorem

At the beginning of Section 1.3 we claimed to be geometers at heart, yet here we are slowly sinking into a morass of technical detail. To escape (temporarily at least; the properties of abstract topological spaces will be examined in more detail in the next chapter) we return to our theory of surfaces.

We shall restrict ourselves to a rather nice class of surfaces, and consider only those which have no boundary and which are in some sense closed up on themselves: in addition we ask that our surfaces be connected, i.e., consist of a single piece. The sphere, the torus, and the Klein bottle are the sort of surfaces that we have in mind; the cylinder and the Möbius strip are ruled out because they have edges. We rule out the whole plane and surfaces such as that represented in Fig. 1.9 as not being 'closed up'. To be precise we are dealing with compact, connected surfaces, but the precise definitions of compactness and connectedness will have to wait until Chapter 3.

The remarkable thing is that if we agree to work only with these so-called 'closed' surfaces, then we can say exactly how many there are, that is, we can *classify* them. Such a classification entails making a list of surfaces so that given an arbitrary closed surface it is homeomorphic to one on the list. In addition, the list should not be too long; in other words no two surfaces on our list should be homeomorphic.

We can construct examples of closed surfaces as follows. Take the ordinary sphere, remove two disjoint discs and then add on a cylinder by identifying its two boundary circles with the boundaries of the holes in the sphere as in Fig. 1.17. This process is called 'adding a handle' to the sphere. By repetition we obtain a sphere with two, three, or any finite number of handles. You should be able to convince yourself that a sphere with one handle is nothing more than (is homeomorphic to) the torus. This process of adding handles gives half the surfaces of our list.

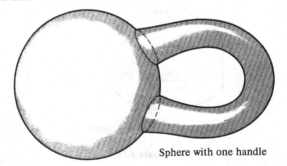

Sphere with one handle

Figure 1.17

The others are unfortunately like the Klein bottle in the sense that they do not admit representatives in euclidean three-dimensional space and are therefore more difficult to imagine. Luckily, the construction of models for these surfaces is an easy process to describe. Begin with the sphere, remove a single disc, and add a Möbius strip in its place. The Möbius strip has after all a single circle as boundary, and all that we are asking is that the points of this boundary circle be identified with those of the boundary circle of the hole in the sphere. One must imagine this identification taking place in some space where there is plenty of room† (euclidean four-dimensional space will do); as noted above, it cannot be realized in three dimensions without having the Möbius strip intersect itself. The resulting closed surface is called the *projective plane*.

For each positive integer n we can form a closed surface by removing n disjoint discs from the sphere and replacing each one by a Möbius strip. When $n = 2$, we recapture the Klein bottle and Fig. 1.18 is an attempt to illustrate

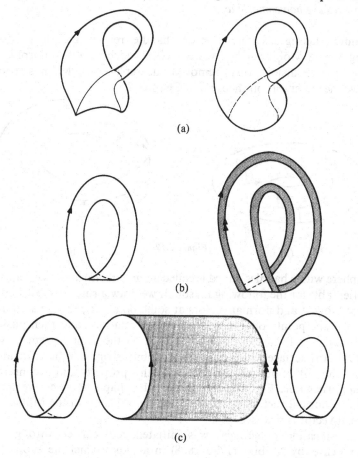

(a)

(b)

(c)

Figure 1.18

† In Chapter 4 we shall explain how to glue two topological spaces together in order to form a new space, without relying in any way on models of the spaces in $\mathbb{E}^3$ or $\mathbb{E}^4$.

why this is so. Slicing the usual picture of the Klein bottle in two in the plane of the paper, and removing the self intersections of the two pieces, gives the two Möbius strips of Fig. 1.18a. Take one of these and mark in a strip neighbourhood of its boundary (1.18b); this neighbourhood is homeomorphic to a cylinder. Remove the cylinder (1.18c), leaving a slightly smaller Möbius strip and remember that a cylinder is homeomorphic to a sphere with two disjoint discs removed. So the usual description of the Klein bottle agrees with our construction when $n = 2$.

(1.5) Classification theorem. *Any closed surface is homeomorphic either to the sphere, or to the sphere with a finite number of handles added, or to the sphere with a finite number of discs removed and replaced by Möbius strips. No two of these surfaces are homeomorphic.*

For example, taking a sphere with one handle, removing a single disc, and replacing it by a Möbius strip gives a surface which is homeomorphic to the sphere with three disjoint discs removed and replaced by Möbius strips. We will prove the classification theorem in Chapter 7.

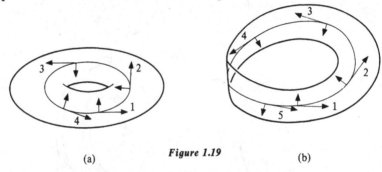

(a) **Figure 1.19** (b)

The sphere with n handles added is called an *orientable* surface of genus n. We call it orientable for the following reason. If we draw a smooth closed curve on it, choose tangent and normal vectors at some point (i.e., choose a coordinate system near the point – often called a local orientation), and then push these vectors once round the curve we come back to the same system of vectors (Fig. 1.19a). Any surface which contains a Möbius strip, and therefore all those on the second half of our list, cannot satisfy this property and is consequently called nonorientable. Figure 1.19b shows what happens when we push the tangent and normal vectors once round the central circle of the Möbius strip – the normal vector is reversed.

The classification of surfaces was initiated, and carried through in the orientable case, by Möbius (1790–1868) in a paper which he submitted for consideration for the Grand Prix de Mathématiques of the Paris Academy of Sciences. He was 71 at the time. The jury did not consider any of the manuscripts received as being worthy of the prize, and Möbius' work finally appeared as just another mathematical paper.

1.6 Topological invariants

We should say at once that we have no hope of classifying all topological spaces. However, we would like to develop ways of deciding whether or not two concrete spaces, such as two surfaces, are homeomorphic.

Showing that two spaces are homeomorphic is a geometrical problem, involving the construction of a specific homeomorphism between them. The techniques used vary with the problem. We have already given an example (at least in outline) in showing that the Klein bottle is homeomorphic to the sphere with two disjoint discs replaced by Möbius strips.

Attempting to prove that two spaces are not homeomorphic to one another is a problem of an entirely different nature. We cannot possibly examine each function between the two spaces individually and check that it is not a homeomorphism. Instead we look for 'topological invariants' of spaces: an invariant may be a geometrical property of the space, a number like the Euler number defined for the space, or an algebraic system such as a group or a ring constructed from the space. The important thing is that the invariant be preserved by a homeomorphism – hence its name. If we suspect that two spaces are not homeomorphic, we may be able to confirm our suspicion by computing some suitable invariant and showing that we obtain different answers. We give two examples below.

In Chapter 3 we shall introduce the notion of connectedness: roughly speaking, a space is connected if it is all in one piece. This idea can be made quite precise, and it will be no surprise to us to find that the property of being connected is preserved if we apply a homeomorphism to a space, i.e., connectedness is a topological invariant. The plane $\mathbb{E}^2$ is an example of a connected space; so is the line $\mathbb{E}^1$. However, if we remove the origin from $\mathbb{E}^1$ the space falls into two pieces (corresponding to the positive and negative real numbers) and we have an example of a space which is not connected. Suppose now that we have a homeomorphism $h:\mathbb{E}^1 \to \mathbb{E}^2$. It will induce a homeomorphism from $\mathbb{E}^1 - \{0\}$ to $\mathbb{E}^2 - \{h(0)\}$. But $\mathbb{E}^2$ with a single point removed is a connected space (it is all in one piece) whereas $\mathbb{E}^1 - \{0\}$ is not connected. We conclude that $\mathbb{E}^1$ and $\mathbb{E}^2$ are not homeomorphic.

As a second example, we consider a construction due to Poincaré which will be the subject of Chapter 5. The idea is to assign a group to each topological space in such a way that *homeomorphic spaces have isomorphic groups*. If we want to distinguish between two spaces, we can try to solve the problem *algebraically* by first computing their groups and then looking to see whether or not the groups are isomorphic. If the groups are not isomorphic then the spaces are different (not homeomorphic). Of course, we may be unlucky and wind up with isomorphic groups, in which case we must look for a more delicate invariant to separate the two spaces.

Consider the two spaces shown in Fig. 1.20. We would not expect them to be homeomorphic, after all the annulus has a hole through it and the disc does not.

This hole is represented very well by the loop α of Fig. 1.21. It is the hole which prevents us from continuously shrinking α to a point without leaving the annulus, whereas in a disc any loop can be continuously shrunk to a point.

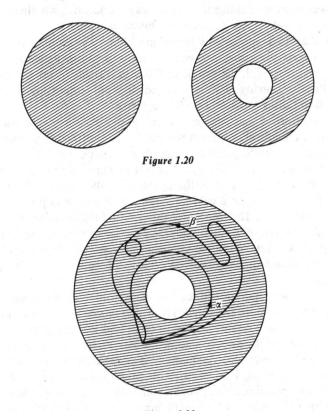

Figure 1.20

Figure 1.21

Poincaré's construction uses loops like α to produce a group, the so-called fundamental group of the annulus: this group will pick out the fact that the annulus has a hole.

A loop such as α will give rise to a nontrivial element of the fundamental group. Looking again at the annulus, the loop β is for our purposes of hole recognition just as good as α, as it can be continuously deformed into α without crossing the hole. This suggests that β should represent the same element of the fundamental group as α. Working with loops which begin and end at a particular point means that there is a natural way of multiplying loops together. One should think of the product α . β of two loops as being the composite loop obtained by first going round the loop α, then going round β. The loops themselves do not form a group under this multiplication, but if we agree to identify loops when one can be continuously deformed into the other (without moving their endpoints), then we do get a group from the resulting equivalence classes of loops.

The above discussion can be made quite precise. Mathematically, a loop in a topological space X is nothing more than a continuous function $\alpha : C \to X$, where C denotes the unit circle in the complex plane, and we say that the loop begins and ends at the point p of X if $\alpha(1) = p$. The arrows on the loops in our illustrations indicate the direction of increasing θ, where we parametrize C as $\{e^{i\theta} \mid 0 \leqslant \theta \leqslant 2\pi\}$. Reversing an arrow produces a different loop and corresponds to taking the inverse of the appropriate element in the fundamental group. Perhaps the simplest possible loop is the function which sends all of C to the point p, and it is this loop which represents the identity element of the fundamental group.

The fundamental group of a disc is the trivial group, since any loop can be continuously shrunk to a point – we leave the technicalities of defining a continuous deformation to Chapter 5. For the annulus we obtain the infinite cyclic group of integers. Loops representing 0, -1, and $+2$ are shown in Fig. 1.22.

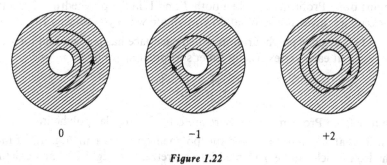

$$0 \qquad\qquad -1 \qquad\qquad +2$$

Figure 1.22

It is not hard to imagine that homeomorphic spaces will have isomorphic fundamental groups. After all, if $\alpha : C \to X$ is a loop in X, and $h : X \to Y$ a homeomorphism, then $h\alpha : C \to Y$ defines a loop in Y; continuous deformations also carry over via a homeomorphism. We conclude that the disc and the annulus are not homeomorphic.

Perhaps the best way for us to complete our introduction, and at the same time to capture the flavour of later chapters, is to list some problems (three from geometry, one from algebra) which we shall use the fundamental group to help solve.

Classification of surfaces. No two surfaces on the list given in theorem (1.5) have isomorphic fundamental groups, so these surfaces are all topologically distinct.

Jordan separation theorem. Any simple closed curve in the plane divides the plane into two pieces.

21

Brouwer fixed-point theorem. Any continuous function from a disc to itself leaves at least one point fixed.

Nielsen–Schreier theorem. A subgroup of a free group is always free.

Problems

1. Prove that $v(T) - e(T) = 1$ for any tree T.

2. Even better, show that $v(\Gamma) - e(\Gamma) \leqslant 1$ for any graph Γ, with equality precisely when Γ is a tree.

3. Show that inside any graph we can always find a tree which contains all the vertices.

4. Find a tree in the polyhedron of Fig. 1.3 which contains all the vertices. Construct the dual graph Γ and show that Γ contains loops.

5. Having done Problem 4, thicken both T and Γ in the polyhedron. T is a tree, so thickening it gives a disc. What do you obtain when you thicken Γ?

6. Let P be a regular polyhedron in which each face has p edges and for which q faces meet at each vertex. Using Euler's formula prove that

$$\frac{1}{p} + \frac{1}{q} = \frac{1}{2} + \frac{1}{e}.$$

7. Deduce from Problem 6 that there are only five regular polyhedra.

8. Check that $v - e + f = 0$ for the polyhedron shown in Fig. 1.3. Find a polyhedron which can be deformed into a pretzel (see Fig. 1.23c) and calculate its Euler number.

9. Borrow a tennis ball and observe that its surface is marked out as the union of two discs which meet along their boundaries.

10. Find a homeomorphism from the real line to the open interval (0,1). Show that any two open intervals are homeomorphic.

11. Imagine all the spaces shown in Fig. 1.23 to be made of rubber. For each pair of spaces X, Y, convince yourself that X can be continuously deformed into Y.

(a)

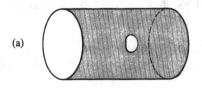
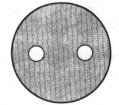

X = Cylinder with a puncture Y = Disc with two punctures

Figure 1.23

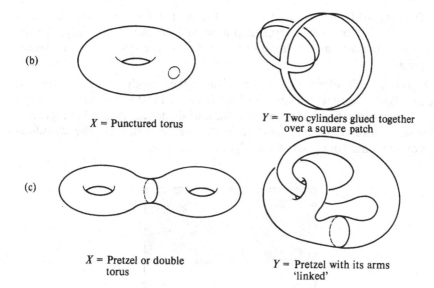

(b)

X = Punctured torus

Y = Two cylinders glued together over a square patch

(c)

X = Pretzel or double torus

Y = Pretzel with its arms 'linked'

12. 'Stereographic projection' π from the sphere minus the north pole to the plane is shown in Fig. 1.24. Work out a formula for π and check that π is a homeomorphism.

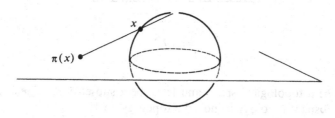

Figure 1.24

Notice that π provides us with a homeomorphism from the sphere with the north and south poles removed to the plane minus the origin.

13. Let x and y be points on the sphere. Find a homeomorphism of the sphere with itself which takes x to y. Work the same problem with the sphere replaced by the plane and by the torus.

14. Make a Möbius strip out of a rectangle of paper and cut it along its central circle. What is the result?

15. Cut a Möbius strip along the circle which lies halfway between the boundary of the strip and the central circle. Do the same for the circle which lies one-third of the way in from the boundary. What are the resulting spaces?

16. Now take a strip which has one full twist in it, cut along its central circle and see what happens.

17. Define $f:[0,1)\to C$ by $f(x)=e^{2\pi ix}$. Prove that f is one–one, onto, and continuous. Find a point $x\in[0,1)$ and a neighbourhood N of x in $[0,1)$ such that $f(N)$ is not a neighbourhood of $f(x)$ in C. Deduce that f is not a homeomorphism.

18. If you had difficulty with Problem 11(b), make a model of a torus minus a disc as follows. Begin with a square whose sides are to be identified in the usual way to give a torus (Fig. 1.25). Note that the four shaded areas together represent a disc in the torus. Cut these areas out of the square, then make the identifications on the remaining parts of the edges of the square.

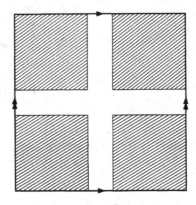

Figure 1.25

19. Let X be a topological space and let Y be a subset of X. Check that the so-called subspace topology is indeed a topology on Y.

20. Prove that the radial projection shown in Fig. 1.8 is a homeomorphism from the surface of the tetrahedron to the sphere. (Both spaces are assumed to have the subspace topology from $\mathbb{E}^3$.)

21. Let C denote the unit circle in the complex plane and D the disc which it bounds. Given two points $x,y\in D-C$, find a homeomorphism from D to D which interchanges x and y and leaves all the points of C fixed.

22. With C, D as above, define $h:D-C\to D-C$ by

$$h(0)=0$$

$$h(r\,e^{i\theta})=r\exp\left[i\left(\theta+\frac{2\pi r}{1-r}\right)\right]$$

Show that h is a homeomorphism, but that h cannot be extended to a homeomorphism from D to D. Draw a picture which shows the effect of h on a diameter of D.

23. Using the intuitive notion of connectedness, argue that a circle and a circle with a spike attached cannot be homeomorphic (Fig. 1.26.)

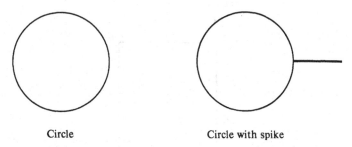

Circle   Circle with spike

Figure 1.26

24. Let X, Y be the subspaces of the plane shown in Fig. 1.27. Under the assumption that any homeomorphism from the annulus to itself must send the points of the two boundary circles among themselves,† argue that X and Y cannot be homeomorphic.

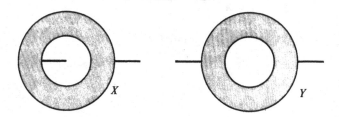

Figure 1.27

25. With X and Y as above, consider the following two subspaces of $\mathbb{E}^3$:

$$X \times [0,1] = \{(x,y,z) \,|\, (x,y) \in X, \quad 0 \leqslant z \leqslant 1\},$$
$$Y \times [0,1] = \{(x,y,z) \,|\, (x,y) \in Y, \quad 0 \leqslant z \leqslant 1\}.$$

Convince yourself that if these spaces are made of rubber then they can be deformed into one another, and hence that they are homeomorphic.

26. Assuming you have done Problem 14, show that identifying diametrically opposite points on one of the boundary circles of the cylinder leads to the Möbius strip.

---

† This is not easy to verify: for a proof see theorem (5.24).

27. Make a model for a Klein bottle as shown in Fig. 1.28. Cut along the line CD, then identify the two lines labelled AB. Inspect the result and deduce that the Klein bottle is made up of two Möbius strips which have a common boundary circle.

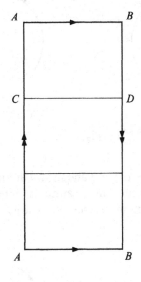

Figure 1.28

2. Continuity

*Geometry formerly was the chief borrower from arithmetic
and algebra, but it has since repaid its obligations with
abundant usury; and if I were asked to name, in one
word, the pole star round which the mathematical
firmament revolves, the central idea which pervades the
whole corpus of mathematical doctrine, I should point to
Continuity as contained in our notions of space, and say,
it is this, it is this!*

J. J. Sylvester

2.1 Open and closed sets

The definition of a topological space given in Chapter 1 fits quite well our
intuitive idea of what a space ought to be. Unfortunately, it is not terribly
convenient to work with, and our first job is to produce an equivalent, more
manageable, set of axioms.

Let X be a topological space and call a subset O of X *open* if it is a neighbourhood of each of its points. Notice that the union of any collection of open sets is
open by axiom (c) of definition (1.3), and the intersection of any *finite* number
of open sets is open by axiom (b). The whole space X is an open set, as is the
empty set $\varnothing$. Also, given a neighbourhood N of a point x, axiom (d) tells us
that the interior of N is an open set which lies inside N and which contains x.

In $\mathbb{E}^3$ a set is open if each of its points can be surrounded by a ball which lies
entirely inside the set. For example, the half-space defined by the inequality
$z > 0$ is open, as is the set of points whose coordinates satisfy $x^2 + y^2 + z^2 < 1$.
On the other hand, the half-space defined by $z \geqslant 0$ is not open because any
ball, however small, which surrounds a point of the (x,y) plane must dip down
into the lower half-space given by $z < 0$. The intersection of an infinite collection of open sets need not be open, for example if we intersect the sets

$$\left\{ (x,y,z) \mid x^2 + y^2 + z^2 < \frac{1}{n} \right\} \qquad n = 1,2,3,\ldots$$

we obtain the origin in $\mathbb{E}^3$, which is not open.

We shall now try to work in the opposite direction, starting from the idea of
an open set, then building up a collection of neighbourhoods for each point.
Suppose then we have a set X together with a nonempty collection of subsets of
X, which we call open sets, such that any union of open sets is itself open, any

finite intersection of open sets is open, and both the whole set and the empty set are open. Given a point x of X, we shall call a subset N of x a *neighbourhood of* x if we can find an open set O such that $x \in O \subseteq N$.

We claim that this definition of neighbourhood makes X into a topological space. Each point has at least one neighbourhood, namely the whole set X, and axioms (a) and (c) of definition (1.3) clearly hold. If N_1, N_2 are neighbourhoods of x we can find open sets O_1, O_2 such that $x \in O_1 \subseteq N_1$ and $x \in O_2 \subseteq N_2$, giving $x \in O_1 \cap O_2 \subseteq N_1 \cap N_2$. But $O_1 \cap O_2$ is open, therefore $N_1 \cap N_2$ is a neighbourhood of x, and we have verified axiom (b). Finally, suppose N is a neighbourhood of x and let $\mathring{N}$ denote the set of points z such that N is a neighbood of z. Choose an open set O such that $x \in O \subseteq N$. Now O, being open, is a neighbourhood of each of its points, so O is contained in $\mathring{N}$. Therefore, $\mathring{N}$ is a neighbourhood of x as required for axiom (d).

Suppose we go full circle. In other words, we start with a collection of so-called open sets, construct a topological space X using them, then look at the open sets of this space. Do the two notions of 'open' coincide? The answer is yes. For if O is one of the original open sets, then it is by definition a neighbourhood of each of its points in X, and therefore an open set of X. Conversely, if U is an open set of X it is a neighbourhood of each of its points. So given $x \in U$ we can find one of the original open sets, say O_x, such that $x \in O_x \subseteq U$. But then $U = \bigcup \{O_x \mid x \in U\}$, and is therefore open in the original sense because any union of open sets is open. We leave the reader to check out the other possibility, namely, if we begin with a topological space, introduce the notion of an open set, then construct a family of neighbourhoods for each point using these open sets, the neighbourhoods which result are precisely those of the original space.

The above discussion means we are justified in rephrasing our definition of a topological space in terms of open sets.

(2.1) Definition. *A topology on a set* X *is a nonempty collection of subsets of* X, *called open sets, such that any union of open sets is open, any finite intersection of open sets is open, and both* X *and the empty set are open. A set together with a topology on it is called a topological space.*

This is the definition we shall adopt from now on.

The open sets of the 'usual' topology on $\mathbb{E}^n$ are characterized as follows. A set U is open if given $x \in U$ we can always find a positive real number ε such that the ball with centre x and radius ε lies entirely in U. Whenever we refer to $\mathbb{E}^n$ we shall have this topology in mind.

If we have a topological space X and a subset Y of X, the open sets of the *subspace* or *induced* topology on Y are obtained simply by intersecting all of the open sets of X with Y. In other words, a subset U of Y is open in the subspace topology if we can find an open set O of X such that $U = O \cap Y$. Any subset of a euclidean space picks up a topology from the surrounding space in this way. Whenever we refer to a *subspace* Y of a topological space X, we shall mean that Y is a subset of X and has the subspace topology.

A rather extreme topology is the *discrete topology* on X. Here every subset of X is an open set. This is the largest possible topology on a given set X. (If one topology contains all the open sets of another, we say it is 'larger' than the other.) If X has the discrete topology we call it a discrete space. For example, if we take the set of points of $\mathbb{E}^n$ which have integer coordinates, and give it the subspace topology, the result is a discrete space.

A subset of a topological space is *closed* if its complement is open. Think of subsets of the plane such as the unit circle, the unit disc (points whose co-ordinates satisfy $x^2 + y^2 \leqslant 1$), the graph of the function $y = e^x$, or the set of points (x,y) such that $x \geqslant y^2$. All these sets are closed. Still working in $\mathbb{E}^2$, consider the set A whose points (x,y) satisfy $x \geqslant 0$ and $y > 0$. This set A is not closed, because the x axis lies in its complement, yet any ball with centre on the positive part of the x axis must meet A. Notice that A is not open either. So sets may be neither open nor closed. They can equally well be both open and closed. For example, take the space X whose points are those points (x,y) of $\mathbb{E}^2$ such that $x \geqslant 1$ or $x \leqslant -1$, and whose topology is that induced from $\mathbb{E}^2$. The subset of X consisting of those points with positive first coordinate is both open and closed in X. (Though of course it is not open in $\mathbb{E}^2$.) We note that the inter-section of any family of closed sets is closed, as is the union of any *finite* family of closed sets. To prove these statements one simply applies the De Morgan formulae.

We can characterize closed sets very nicely as follows. Let A be a subset of a topological space X and call a point p of X a *limit point* (or accumulation point) of A if every neighbourhood of p contains at least one point of $A - \{p\}$. Such a point may or may not be in A as the following examples show.

Examples.
1. Take X to be the real line $\mathbb{R}$ (the usual name and notation for $\mathbb{E}^1$), and let A consist of the points $1/n, n = 1,2,\ldots$. Then A has exactly one limit point, namely the origin.
2. Again with X as the real line, take $A = [0,1)$. Then each point of A is a limit point of A, and in addition 1 is a limit point of A.
3. Let X be $\mathbb{E}^3$ and let A consist of those points all of whose coordinates are rational. Then every point of $\mathbb{E}^3$ is a limit point of A.
4. At the other extreme, let $A \subseteq \mathbb{E}^3$ be the set of points which have integer coordinates. Then A does not have any limit points.
5. Take X to be the set of all real numbers with the so called *finite-complement topology*. Here a set is open if its complement is finite or all of X. If we now take A to be an infinite subset of X (say the set of all integers), then every point of X is a limit point of A. On the other hand a finite subset of X has no limit points in this topology.

(2.2) Theorem. *A set is closed if and only if it contains all its limit points.*

Proof. If A is closed, its complement $X - A$ is open. Since an open set is a neighbourhood of each of its points, no point of $X - A$ can be a limit point of A. Therefore A contains all its limit points. Conversely, suppose A contains all its limit points and let $x \in X - A$. Since x is not a limit point of A we can find a neighbourhood N of x which does not meet A. So N is inside $X - A$, showing $X - A$ to be a neighbourhood of each of its points and consequently open. Therefore A is closed.

The union of A and all its limit points is called the *closure* of A and is written $\bar{A}$.

(2.3) Theorem. *The closure of* A *is the smallest closed set containing* A, *in other words the intersection of all closed sets which contain* A.

Proof. We first observe that $\bar{A}$ is indeed a closed set. For if $x \in X - \bar{A}$, we can find an open neighbourhood U of x which does not contain any points of A. Since an open set is a neighbourhood of each of its points, U cannot contain any of the limit points of A either. Therefore we have an open set U such that $x \in U \subseteq X - \bar{A}$. Consequently $X - \bar{A}$ is a neighbourhood of each of its points and must be open. Now let B be a closed set which contains A. Then every limit point of A is a limit point of B and therefore must lie in B since B is closed. This gives $\bar{A} \subseteq B$. Since $\bar{A}$ is closed, contains A, and is contained in every closed set which contains A, it must be the intersection of all such sets.

(2.4) Corollary. *A set is closed if and only if it is equal to its closure.*

A set whose closure is the whole space is said to be *dense* in the space. This is the case in example 3 above. A dense set meets every nonempty open subset of the space.

The *interior* of a set A, usually written $\mathring{A}$, is the union of all open sets contained in A. One readily checks that a point x lies in the interior of A if and only if A is a neighbourhood of x. An open set is its own interior; if we work in $\mathbb{E}^2$ and use D to denote the unit disc consisting of points (x,y) such that $x^2 + y^2 \leqslant 1$, the interior of D is $D - C$, where C stands for the unit circle; the circle C has empty interior because the only open set of the plane contained in C is the empty set.

One other useful notion is that of the *frontier* of a set. We define the frontier of A to be the intersection of the closure of A with the closure of $X - A$. An equivalent definition is to take those points of X which do not belong to the interior of A nor to the interior of $X - A$. For example, in the plane, the unit disc D, its interior $\mathring{D}$, and the unit circle C all have the same frontier, namely C. The frontier of the set of points in $\mathbb{E}^3$ which have rational coordinates is all of $\mathbb{E}^3$, so the frontier of a set can be the whole space.

Suppose we have a topology on a set X, and a collection β of open sets such that every open set is a union of members of β. Then β is called a *base* for the topology and elements of β are called *basic open sets*. An equivalent formulation is to ask that given a point $x \in X$, and a neighbourhood N of x, there is always an element B of β such that $x \in B \subseteq \beta$. A good example is provided by the topology of the real line, where the set of all open intervals is a base. The set of open intervals which have rational endpoints is a smaller base. (Notice that this second base is countable.)

It can be useful to describe a topology on a set by specifying a base for the topology. For this reason we would like to be able to decide when a given collection of subsets of a set X is a base for some topology on X.

(2.5) Theorem. *Let β be a nonempty collection of subsets of a set* X. *If the intersection of any finite number of members of β is always in β, and if $\bigcup \beta = X$, then β is a base for a topology on* X.

Proof. Take the obvious candidate, namely the collection of all unions of members of β as the open sets, then check the requirements for a topology.

Problems

1. Verify each of the following for arbitrary subsets A, B of a space X:
(a) $\overline{A \cup B} = \overline{A} \cup \overline{B}$; (b) $\overline{A \cap B} \subseteq \overline{A} \cap \overline{B}$; (c) $\overline{\overline{A}} = \overline{A}$;
(d) $(A \cup B)^\circ \supseteq \mathring{A} \cup \mathring{B}$; (e) $(A \cap B)^\circ = \mathring{A} \cap \mathring{B}$; (f) $(\mathring{A})^\circ = \mathring{A}$.
Show that equality need not hold in (b) and (d).

2. Find a family of closed subsets of the real line whose union is not closed.

3. Specify the interior, closure, and frontier of each of the following subsets of the plane:
(a) $\{(x,y) \mid 1 < x^2 + y^2 \leqslant 2\}$; (b) $\mathbb{E}^2$ with both axes removed;
(c) $\mathbb{E}^2 - \{(x, \sin(1/x)) \mid x > 0\}$.

4. Find all the limit points of the following subsets of the real line:
(a) $\{(1/m) + (1/n) \mid m,n = 1,2,\ldots\}$; (b) $\{(1/n) \sin n \mid n = 1,2\ldots\}$.

5. If A is a dense subset of a space X, and if O is open in X, show that $O \subseteq \overline{A \cap O}$.

6. If Y is a subspace of X, and Z a subspace of Y, prove that Z is a subspace of X.

7. Suppose Y is a subspace of X. Show that a subset of Y is closed in Y if it is the intersection of Y with a closed set in X. If A is a subset of Y, show that we get the same answer whether we take the closure of A in Y, or intersect Y with the closure of A in X.

8. Let Y be a subspace of X. Given $A \subseteq Y$, write $\mathring{A}_Y$ for the interior of A in Y, and $\mathring{A}_X$ for the interior of A in X. Prove that $\mathring{A}_X \subseteq \mathring{A}_Y$, and give an example to show the two may not be equal.

9. Let Y be a subspace of X. If A is open (closed) in Y, and if Y is open (closed) in X, show that A is open (closed) in X.

10. Show that the frontier of a set always contains the frontier of its interior. How does the frontier of $A \cup B$ relate to the frontiers of A and B?

11. Let X be the set of real numbers and β the family of all subsets of the form $\{x \mid a \leqslant x < b$ where $a < b\}$. Prove that β is a base for a topology on X and that in this topology each member of β is both open and closed. Show that this topology does not have a countable base.

12. Show that if X has a countable base for its topology, then X contains a countable dense subset. A space whose topology has a countable base is called a *second countable* space. A space which contains a countable dense subset is said to be *separable*.

2.2 Continuous functions

The notion of continuity is particularly easy to formulate in terms of open sets. Let X and Y be topological spaces.

(2.6) Theorem. *A function from* X *to* Y *is continuous if and only if the inverse image of each open set of* Y *is open in* X.

Proof. Recall the definition of continuity given in Chapter 1. A function $f : X \to Y$ is continuous if for each point x of X and each neighbourhood N of $f(x)$ in Y the set $f^{-1}(N)$ is a neighbourhood of x in X. Now if f is continuous and if O is an open subset of Y, then O is a neighbourhood of each of its points and therefore $f^{-1}(O)$ must be a neighbourhood of each of its points in X. We conclude that $f^{-1}(O)$ is an open set in X. The converse implication is left to the reader.

A continuous function is very often called a *map* for short.

(2.7) Theorem. *The composition of two maps is a map.*

Proof. Suppose $f : X \to Y$, $g : Y \to Z$ are continuous; let O be an open set in Z and notice that $(g \circ f)^{-1}(O) = f^{-1}g^{-1}(O)$. Now $g^{-1}(O)$ is open in Y because g is continuous, so $f^{-1}g^{-1}(O)$ must be open in X by the continuity of f. Therefore $g \circ f$ is continuous.

(2.8) Theorem. *Suppose* f : X $\to$ Y *is continuous, and let* A $\subseteq$ X *have the subspace topology. Then the restriction* f$\mid$A : A $\to$ Y *is continuous.*

Proof. Let O be an open set in Y and notice that $(f \mid A)^{-1}(O) = A \cap f^{-1}(O)$. Since f is continuous, $f^{-1}(O)$ is open in X. Therefore $(f \mid A)^{-1}(O)$ is open in the subspace topology on A, and the continuity of $f \mid A$ follows from theorem (2.6).

The map from X to X which sends each point x to itself is called the *identity map of* X and written 1_X. If we restrict 1_X to a subspace A of X we obtain the *inclusion map* $i : A \to X$.

(2·9) Theorem. *The following are equivalent:*
(a) $f : X \to Y$ *is a map.*
(b) *If* β *is a base for the topology of* Y, *the inverse image of every member of* β *is open in* X.
(c) $f(\bar{A}) \subseteq \overline{f(A)}$ *for any subset* A *of* X.
(d) $\overline{f^{-1}(B)} \subseteq f^{-1}(\bar{B})$ *for any subset* B *of* Y.
(e) *The inverse image of each closed set in* Y *is closed in* X.

Proof. The most efficient way of dealing with this is to verify the five implications (a) $\Rightarrow$ (b) $\Rightarrow$ (c) $\Rightarrow$ (d) $\Rightarrow$ (e) $\Rightarrow$ (a). We shall deal with two of them and leave the other three to the reader. Consider (b) $\Rightarrow$ (c). Let A be a subset of X. Certainly every point of $f(A)$ lies in $\overline{f(A)}$, therefore we must show that if $x \in \bar{A} - A$, and $f(x) \notin f(A)$, the point $f(x)$ is a limit point of $f(A)$. If N is a neighbourhood of $f(x)$ in Y we can find a basic open set B in β such that $f(x) \in B \subseteq N$. Assuming (b), the set $f^{-1}(B)$ is open in X and is therefore a neighbourhood of x. But x is a limit point of A, which means that $f^{-1}(B)$ must contain a point of A. So B, and therefore N, contains a point of $f(A)$ as required. To prove (d) $\Rightarrow$ (e) we note that if B is a closed subset of Y then $\bar{B} = B$. But if we assume (d), we have $\overline{f^{-1}(B)} \subseteq f^{-1}(\bar{B}) = f^{-1}(B)$. So $f^{-1}(B)$ is closed in X.

Example. Let C denote the unit circle in the complex plane, taken with the subspace topology, and give the interval $[0,1)$ the induced topology from the real line. Define $f : [0,1) \to C$ by $f(x) = e^{2\pi i x}$. It is easy to see that f is continuous. We can take the set of all open segments of the circle as a base for the topology on C. Now if S is such a segment, and if S does not contain the complex number 1, then $f^{-1}(S)$ is just an open interval of the form (a,b) where $0 < a < b < 1$. So $f^{-1}(S)$ is open in $[0,1)$. If S does happen to contain 1 (as in Fig. 2.1) then

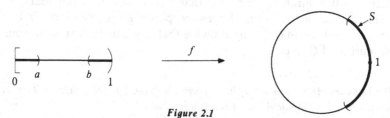

Figure 2.1

$f^{-1}(S)$ has the form $[0,a) \cup (b,1)$ where $0 < a < b < 1$. This is open in $[0,1)$ because it is the intersection of the open set $(-1,a) \cup (b,1)$ of the real line with $[0,1)$. Part (b) of theorem (2.9) now establishes the continuity of f. Our function is clearly one–one and onto. However, its inverse is not continuous. To see this we need only produce an open set O of $[0,1)$ such that $(f^{-1})^{-1}(O) = f(O)$

is not open in C. Take O to be the interval $[0,\frac{1}{2})$; this is open in $[0,1)$, but its image under the exponential map consists of those complex numbers z in C for which $0 \leqslant \arg z < \pi$, and this set is not open in C.

A *homeomorphism* $h : X \to Y$ is a function which is continuous, one–one, and onto, and which has continuous inverse. From theorem (2.6) we see that a set O is open in X if and only if $h(O)$ is open in Y. Therefore, h induces a one–one onto correspondence between the topologies of X and Y, justifying our assertion that X and Y should be thought of as the same topological space.

Example. Let S^n denote the n-dimensional sphere whose points are those of $\mathbb{E}^{n+1}$ which have distance 1 from the origin, taken with the subspace topology. We claim that removing a single point from S^n gives a space homeomorphic to $\mathbb{E}^n$. Which point we remove is irrelevant because we can rotate any point of S^n into any other; for convenience we choose to remove the point $p = (0,\ldots,0,1)$. Now the set of points of $\mathbb{E}^{n+1}$ which have zero as their final coordinate, when given the induced topology, is clearly homeomorphic to $\mathbb{E}^n$. We define a function $h : S^n - \{p\} \to \mathbb{E}^n$, called *stereographic projection*, as follows. If $x \in S^n - \{p\}$, then $h(x)$ is the point of intersection of $\mathbb{E}^n$ and the straight line determined by x and p. (For a picture when $n = 2$ see Fig. 1.24.)

Clearly h is one–one and onto. If O is an open set in $\mathbb{E}^n$, we form a new set U in $\mathbb{E}^{n+1}$ whose points are those which lie on the half lines which start at p and pass through points of O, except the point p itself which we rule out. One readily checks that U is open in $\mathbb{E}^{n+1}$. But $h^{-1}(O)$ is precisely the intersection of U with $S^n - \{p\}$, and therefore $h^{-1}(O)$ is open in $S^n - \{p\}$. This establishes the continuity of h, and a precisely similar argument deals with h^{-1}. Therefore h is a homeomorphism.

We end this section with a couple of results which will be needed in the chapter on surfaces. By a *disc* we shall mean any space homeomorphic to the closed unit disc D in $\mathbb{E}^2$. As usual, C stands for the unit circle. If A is a disc, and if $h : A \to D$ is a homeomorphism, then $h^{-1}(C)$ is called the boundary of A and is written ∂A. It is intuitively obvious that this definition of boundary is independent of the choice of the homeomorphism h. We shall verify this fact rigorously in theorem (5.24) by showing that any homeomorphism from D to itself must send C to C.

(2.10) Lemma. *Any homeomorphism from the boundary of a disc to itself can be extended to a homeomorphism of the whole disc.*

Proof. Let A be a disc and choose a homeomorphism $h : A \to D$. Given a homeomorphism $g : \partial A \to \partial A$, we can easily extend $h g h^{-1} : C \to C$ to a homeomorphism of all of D as follows. Send 0 to 0, and if $x \in D - \{0\}$ send x to the point $\| x \| h g h^{-1}(x/\| x \|)$. In other words extend conically. If we call this extension f, then $h^{-1} f h$ extends g to a homeomorphism of all of A as required.

34

(2.11) Lemma. *Let A and B be discs which intersect along their boundaries in an arc. Then A ∪ B is a disc.*

Proof. Let γ denote the arc $A \cap B$, and use α, β for the complementary arcs in the boundaries of A and B (Fig. 2.2). We construct a homeomorphism from

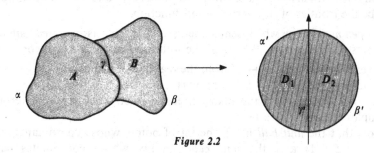

Figure 2.2

$A \cup B$ to D, with the aid of lemma (2.10), as follows. The y axis in the plane divides up D in a particularly nice way, as the union of two discs D_1 and D_2. We label the three arcs which together make up the boundaries of D_1 and D_2 as α', β', γ' as shown in Fig. 2.2. Both α and α' are homeomorphic to the closed unit interval $[0,1]$, so we can find a homeomorphism from α to α'. We first extend this over γ, to give a homeomorphism from $\alpha \cup \gamma$ to $\alpha' \cup \gamma'$ (this much is easy); then over A to give a homeomorphism from A to D_1 which takes γ to γ', using lemma (2.10). Finally, we extend our homeomorphism over β, so that β goes to β', using our common sense, then over B by means of lemma (2.10) again. The result is a homeomorphism from $A \cup B$ to $D_1 \cup D_2 = D$. Therefore $A \cup B$ is a disc.

Problems

13. If $f : \mathbb{R} \to \mathbb{R}$ is a map (i.e., a continuous function), show that the set of points which are left fixed by f is a closed subset of $\mathbb{R}$. If g is a continuous real-valued function on X show that the set $\{x \mid g(x) = 0\}$ is closed.

14. Prove that the function $h(x) = e^x/(1 + e^x)$ is a homeomorphism from the real line to the open interval $(0,1)$.

15. Let $f : \mathbb{E}^1 \to \mathbb{E}^1$ be a map and define its graph $\Gamma_f : \mathbb{E}^1 \to \mathbb{E}^2$ by $\Gamma_f(x) = (x, f(x))$. Show that Γ_f is continuous and that its image (taken with the topology induced from $\mathbb{E}^2$) is homeomorphic to $\mathbb{E}^1$.

16. What topology must X have if every real-valued function defined on X is continuous?

17. Let X denote the set of all real numbers with the finite-complement topology, and define $f : \mathbb{E}^1 \to X$ by $f(x) = x$. Show that f is continuous, but is not a homeomorphism.

35

18. Suppose $X = A_1 \cup A_2 \cup \ldots$, where $A_n \subseteq \mathring{A}_{n+1}$ for each n. If $f : X \to Y$ is a function such that, for each n, $f \mid A_n : A_n \to Y$ is continuous with respect to the induced topology on A_n, show that f is itself continuous.

19. The *characteristic function* of a subset A of a space X is the real-valued function on X which assigns the value 1 to points of A and 0 to all other points. Describe the frontier of A in terms of this function.

20. An *open map* is one which sends open sets to open sets; a *closed map* takes closed sets to closed sets. Which of the following maps are open or closed?

(a) The exponential map $x \mapsto e^{ix}$ from the real line to the circle.
(b) The folding map $f : \mathbb{E}^2 \to \mathbb{E}^2$ given by $f(x,y) = (x, |y|)$.
(c) The map which winds the plane three times on itself given, in terms of complex numbers, by $z \mapsto z^3$.

21. Show that the *unit ball* in $\mathbb{E}^n$ (the set of points whose coordinates satisfy $x_1^2 + \ldots + x_n^2 \leqslant 1$) and the *unit cube* (points whose coordinates satisfy $|x_i| \leqslant 1, 1 \leqslant i \leqslant n$) are homeomorphic if they are both given the subspace topology from $\mathbb{E}^n$.

2.3   A space-filling curve

At the end of the last century Guiseppe Peano made a surprising, and at first sight, paradoxical, discovery. He pointed out the existence of a continuous function defined on a closed interval of the real line which maps the interval *onto* a two-dimensional region in the plane, say onto a square or triangle. Such a function is called a *Peano curve* or *space-filling curve*. One thinks of the image of the interval as a curve which goes through every single point of the two-dimensional region in question.

The existence of space-filling curves shows that a great deal of care is necessary when defining the dimension of a space. Taking the dimension of X to be the least number of continuous parameters needed to specify each point of X is no good. Peano's example shows that the square has dimension 1 under this definition. For a brief discussion of dimension we refer the reader to Chapter 9.

There are many versions of Peano's construction. Here is a simple one which has an equilateral triangle as image. As we might guess, our space-filling curve will be the limit of a sequence of simpler curves which fill out more and more of the triangle as we go along the sequence. Let Δ be an equilateral triangle in the plane whose sides have length one half, and construct a sequence of continuous functions $f_n : [0,1] \to \Delta$ as follows. The first three functions are adequately described by Fig. 2.3, and further members of the sequence are obtained by iterating the procedure shown there. At any particular stage Δ is divided into a number of congruent triangles, and the part of the curve inside each triangle looks precisely like the image of f_1 and joins two vertices of the triangle by a broken line which passes through its centre of gravity. To pass to the next stage we subdivide each triangle into four smaller congruent triangles and insert the more complicated curve which is shown as the image of f_2. As we keep subdividing, the image of f_n fills out more and more of Δ.

Given two points x and y of $\mathbb{E}^2$, we shall use $\| x - y \|$ to denote the distance between them. Suppose $n \geqslant m$, then given $t \in [0,1]$ we can find a triangle which contains both $f_m(t)$ and $f_n(t)$ and whose sides have length $1/2^m$. Therefore $\| f_m(t) - f_n(t) \| \leqslant 1/2^m$ for every value of t in $[0,1]$, which proves that our sequence $\{f_n\}$ is uniformly convergent. Let $f : [0,1] \to \Delta$ denote the limit function. Since each f_n is continuous, so is f.

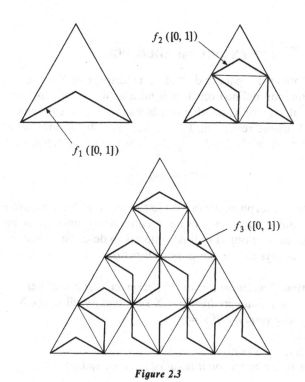

Figure 2.3

We are left to show that the image of f really is all of Δ. First note that, for any n, the image of f_n comes within $1/2^n$ of every point of Δ. Suppose we are given a point x of Δ together with a neighbourhood U of x in $\mathbb{E}^2$. Choose N large enough so that the disc centre x, radius $1/2^{N-1}$, lies inside U, and choose a point t_0 from $[0,1]$ such that $\| x - f_N(t_0) \| \leqslant 1/2^N$. Since $\| f_N(t) - f(t) \| \leqslant 1/2^N$ for every t in $[0,1]$, the triangle inequality gives $\| x - f(t_0) \| \leqslant 1/2^{N-1}$. Therefore $f(t_0)$ must lie inside U. This argument shows every point of Δ to be a limit point of the set $f([0,1])$. But, as we shall see in the next chapter (theorems (3.4) and (3.9)), the image of a continuous function from $[0,1]$ to $\mathbb{E}^2$ must be a closed subset of $\mathbb{E}^2$, and must therefore contain all its limit points. We conclude that the image of our function f is all of Δ.

Problems

22. Find a Peano curve which fills out the unit square in $\mathbb{E}^2$.

23. Find an onto, continuous function from $[0,1]$ to S^2.

24. Can a space-filling curve fill out all of the plane?

25. Can a space-filling curve fill out all of the unit cube in $\mathbb{E}^3$?

26. Do you think that a Peano curve can be one–one? (See theorem (3.7).)

2.4 The Tietze extension theorem

Let X be a topological space and let A be a subspace of X. Given a real-valued continuous function defined on A, it is natural to ask whether or not we can always extend it to all of X. In other words, can we find a real-valued continuous function on X whose restriction to A is the given function? The answer is, in general, no. For example, let $X = [0,1]$, $A = (0,1)$, and define $f:(0,1) \to \mathbb{E}^1$ by

$$f(x) = \log \frac{x}{1 - x}$$

Then f is a homeomorphism from $(0,1)$ to the real line, but f cannot be extended to the closed unit interval because any continuous function defined on $[0,1]$ must be bounded. The object of this section is to describe a particular situation where we can always extend continuous functions.

(2.12) Definition. *A* metric *or* distance function *on a set* X *is a real-valued function* d *defined on the cartesian product* $X \times X$ *such that for all* $x,y,z \in X$:
(a) $d(x,y) \geqslant 0$ *with equality iff* $x = y$;
(b) $d(x,y) = d(y,x)$;
(c) $d(x,y) + d(y,z) \geqslant d(x,z)$.
A set together with a metric on it is called a metric space.

The idea of a metric space is very useful in analysis and the reader may well be familiar with several examples. Any euclidean space, with the usual distance between points, is a metric space, as is the set of all real-valued continuous functions defined on $[0,1]$ with the distance between two functions defined by

$$d(f,g) = \sup_{t \in [0,1]} |f(t) - g(t)|.$$

Any subset of a metric space inherits a metric from the whole space, so a surface in $\mathbb{E}^3$ is a metric space.

A metric on a set gives rise to a topology on the set as follows. Let d be a metric on the set X. Given $x \in X$, the set $\{y \in X \mid d(x,y)\} \leqslant \varepsilon$ is called the *ball of radius* ε, or ε-*ball*, centred at the point x, and is denoted by $B(x,\varepsilon)$. We define a subset O of X to be open if given $x \in O$ we can find a positive real number ε

such that $B(x,\varepsilon)$ is contained in O. The axioms for a topology are easily checked.

Note that different metrics on a set may give the same topology. For example, we can make the underlying set of points of euclidean n-space into a metric space in three different ways as follows. Write $\mathbf{x} = (x_1, x_2, \ldots, x_n)$ for a typical point of $\mathbb{E}^n$ and define:

(a) $d_1(\mathbf{x},\mathbf{y}) = [(x_1 - y_1)^2 + \ldots + (x_n - y_n)^2]^{\frac{1}{2}}$;

(b) $d_2(\mathbf{x},\mathbf{y}) = \max_{1 \leqslant i \leqslant n} |x_i - y_i|$;

(c) $d_3(\mathbf{x},\mathbf{y}) = |x_1 - y_1| + \ldots + |x_n - y_n|$.

Figure 2.4 shows the ball of radius 1, centred at the origin, for each of these three metrics when $n = 2$. To see that d_1 and d_2 give rise to the same topology, we note that inside any disc we can find a square, and conversely inside a square

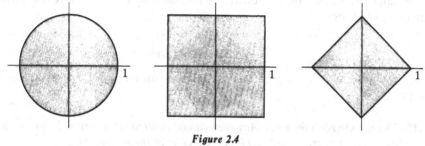

Figure 2.4

we can find a disc. So d_1 and d_2 determine the same open sets. The same remarks hold if we replace the disc or square by the diamond shape of metric d_3. Therefore all three of these metrics give rise to the usual topology on $\mathbb{E}^2$. We leave the reader to work out the general case.

Given two distinct points in a metric space, we can always find disjoint open sets containing them. For if $d(x,y) = \delta > 0$, set $U = \{z \in X \mid d(x,z) < \delta/2\}$ and $V = \{z \in X \mid d(y,z)\} < \delta/2$. Then both U and V are open sets (they are in fact the interiors of $B(x,\delta/2)$ and $B(y,\delta/2)$ respectively), they are disjoint, and of course x lies in U and y lies in V. The set U is usually called the *open ball* with centre x and radius $\delta/2$. A topological space with the property that two distinct points can always be surrounded by disjoint open sets is called a *Hausdorff space*. Not every topological space is Hausdorff; for example, if we give the set of all real numbers the finite-complement topology, then any two nonempty open sets overlap.

If d is a metric on X, and if A is a subset of X, the distance $d(x,A)$ of the point x from A is defined to be the infimum of the numbers $d(x,a)$ where $a \in A$.

(2.13) Lemma. *The real-valued function on* X *defined by* $\mathbf{x} \mapsto d(\mathbf{x},A)$ *is continuous.*

Proof. Let $x \in X$ and let N be a neighbourhood of $d(x,A)$ on the real line. Choose $\varepsilon > 0$ small enough so that the interval $(d(x,A) - \varepsilon, d(x,A) + \varepsilon)$ lies

39

inside N. Let U denote the open ball centre x, radius $\varepsilon/2$, and choose a point $a \in A$ such that $d(x,a) < d(x,A) + \varepsilon/2$. If $z \in U$ we have

$$d(z,A) \leqslant d(z,a) \leqslant d(z,x) + d(x,a) < d(x,A) + \varepsilon$$

By reversing the roles of x and z we also have $d(x,A) < d(z,A) + \varepsilon$. Therefore U is mapped inside $(d(x,A) - \varepsilon, d(x,A) + \varepsilon)$, and hence inside N, by our function, showing that the inverse image of N is a neighbourhood of x in X as required.

(2.14) Lemma. *If* A,B *are disjoint closed subsets of a metric space* X *there is a continuous real-valued function on* X *which takes the value 1 on points of* A, -1 *on points of* B, *and values strictly between* ± 1 *on points of* X $- (A \cup B)$.

Proof. Since A and B are both closed, and are disjoint, the expression $d(x,A) + d(x,B)$ can never be zero (see Problem 27). Therefore we can define a real-valued function f on X by

$$f(x) = \frac{d(x,B) - d(x,A)}{d(x,A) + d(x,B)}$$

Clearly, f takes on the required values, and its continuity follows easily from lemma (2.13).

(2.15) Tietze extension theorem. *Any real-valued continuous function defined on a closed subset of a metric space can be extended over the whole space.*

Proof. Let X be a metric space, C a closed subset, and $f : C \rightarrow \mathbb{E}^1$ a map. To begin with we shall assume that f is bounded; say $|f(x)| \leqslant M$ for all x in C.

Let A_1 consist of those points of C for which $f(x) \geqslant M/3$, and B_1 those for which $f(x) \leqslant -M/3$. Then A_1 and B_1 are obviously disjoint, and they are both closed subsets of X. For example, A_1 is the inverse image of the closed subset $[M/3, \infty)$ of $\mathbb{E}^1$, and is therefore closed in C by the continuity of f. But C is closed in X, and therefore A_1 must be closed in X. A similar argument works for B_1. By lemma (2.14) we can find a map $g_1 : X \rightarrow [-M/3, M/3]$ which takes the value $M/3$ on A_1, $-M/3$ on B_1, and which takes values in $(-M/3, M/3)$ on $X - (A_1 \cup B_1)$. Notice that $|f(x) - g_1(x)| \leqslant 2M/3$ on C.

Now consider the function $f(x) - g_1(x)$ and let A_2 consist of those points of C for which $f(x) - g_1(x) \geqslant 2M/9$, and B_2 those points for which $f(x) - g_1(x) \leqslant -2M/9$. We apply lemma (2.14) a second time to find a map $g_2 : X \rightarrow [-2M/9, 2M/9]$ which takes the value $2M/9$ on A_2, $-2M/9$ on B_2, and values in $(-2M/9, 2M/9)$ on the remaining points of X. If we compute $f(x) - g_1(x) - g_2(x)$, we see that $|f(x) - g_1(x) - g_2(x)| < 4M/9$ on C.

By repeating this process we can construct a sequence of maps $g_n : X \rightarrow [-2^{n-1}M/3^n, 2^{n-1}M/3^n]$ which satisfy:

(a) $|f(x) - g_1(x) - \ldots - g_n(x)| \leqslant 2^n M/3^n$ on C; and

(b) $|g_n(x)| < 2^{n-1}M/3^n$ on $X - C$.

The series $\sum\limits_{n=1}^{\infty} g_n(x)$ converges uniformly on X (by the Weierstrass M-test), so it has a well-defined sum $g(x)$ which is continuous. Also, f and g agree on C by (a). Therefore g extends f to all of X. We note, for use in the unbounded case below, that $|g(x)|$ is bounded by M because

$$|g(x)| \leqslant \sum_{n=1}^{\infty} |g_n(x)| \leqslant M \sum_{n=1}^{\infty} 2^{n-1}/3^n = M$$

and $|g(x)|$ is strictly less than M on $X - C$ by (b).

If the given map f is not bounded, choose a homeomorphism h from the real line to the interval $(-1,1)$ and consider the composition $h \circ f$. This is bounded, and by the above argument we can extend it to a continuous real-valued function g on X, all of whose values lie strictly between -1 and 1. So the composition $h^{-1} \circ g$ is well defined, and by construction it extends f over X. This completes the proof.

We shall make use of the Tietze theorem in Section 5.6.

Problems

27. Show $d(x,A) = 0$ iff x is a point of $\bar{A}$.

28. If A,B are disjoint closed subsets of a metric space, find disjoint open sets U,V such that $A \subseteq U$ and $B \subseteq V$.

29. Show one can define a distance function on an arbitrary set X by $d(x,y) = 1$ if $x \neq y$ and $d(x,x) = 0$. What topology does d give to X?

30. Show that every closed subset of a metric space is the intersection of a countable number of open sets.

31. If A,B are subsets of a metric space, their *distance apart* $d(A,B)$ is the infimum of the numbers $d(x,y)$ where $x \in A$ and $y \in B$. Find two disjoint closed subsets of the plane which are zero distance apart. The *diameter* of A is the supremum of the numbers $d(x,y)$ where $x,y \in A$. Check that both of the closed sets which you have just found have infinite diameter.

32. If A is a closed subset of a metric space X, show that any map $f : A \to E^n$ can be extended over X.

33. Find a map from $E^1 - \{0\}$ to E^1 which cannot be extended over E^1.

34. Let $f : C \to C$ be the identity map of the unit circle in the plane. Extend f to a map from $E^2 - \{0\}$ to C. Would you expect to be able to extend f over all of E^2? (For a precise solution to this latter problem see Section 5.5.)

35. Given a map $f : X \to E^{n+1} - \{0\}$ find a map $g : X \to S^n$ which agrees with f on the set $f^{-1}(S^n)$.

36. If X is a metric space and A closed in X, show that a map $f: A \to S^n$ can always be extended over a *neighbourhood of A*, in other words over a subset of X which is a neighbourhood of each point of A. (Think of S^n as a subspace of $\mathbb{E}^{n+1}$ and extend f to a map of X into $\mathbb{E}^{n+1}$. Now use Problem 35.)

3. Compactness and Connectedness

3.1 Closed bounded subsets of $\mathbb{E}^n$

Those subsets of a euclidean space $\mathbb{E}^n$ which are both closed and bounded† will be of special importance to us. As examples we mention the surfaces described in Chapter 1 and the finite simplicial complexes which we shall construct in Chapter 6 in order to triangulate spaces. We shall show that one can characterize these subsets by a purely topological property, that is to say a property which involves only the topological structure of $\mathbb{E}^n$ and makes no mention of the idea of distance. This property, when formulated for topological spaces in general, is called 'compactness'.

Before giving more details it is convenient to introduce some terminology. Let X be a topological space and let $\mathscr{F}$ be a family of open subsets of X whose union is all of X. Such a family will be called an *open cover* of X. If $\mathscr{F}'$ is a sub-family of $\mathscr{F}$ and if $\bigcup \mathscr{F}' = X$, then $\mathscr{F}'$ is called a *subcover* of $\mathscr{F}$. We give two examples. Let X be the plane and for $\mathscr{F}$ take the collection of all open balls of radius 1 whose centres have integer coordinates. These balls form an open cover of the plane. Notice that if we remove any ball B from $\mathscr{F}$ then the resulting family of balls fails to cover the plane, since its union does not contain the centre of B. Therefore $\mathscr{F}$ has no proper subcover. For our second example we let X be the closed unit interval $[0,1]$ with its usual topology induced from the real line, and take the following family of open subsets of $[0,1]$ for $\mathscr{F}$:

$$[0, 1/10); (1/3, 1]; \quad \text{the sets } (1/(n + 2), 1/n) \text{ where } n \in \mathbb{Z} \text{ and } n \geqslant 2.$$

This open cover is infinite; however, we obviously do not need all of these sets in order to cover the unit interval. We can manage with only a finite number of them, namely

$$[0, 1/10); (1/3, 1]; \quad \text{and } (1/(n + 2), 1/n) \text{ for } 2 \leqslant n \leqslant 9.$$

So this open cover of $[0,1]$ contains a *finite* subcover. In fact, as we shall see in the next section, *any* open cover of $[0,1]$ contains a finite subcover. It is this property which picks out the closed bounded subsets of $\mathbb{E}^n$.

---

† Bounded means contained in some ball which has centre the origin and finite radius.

(3.1) Theorem. *A subset* X *of* $\mathbb{E}^n$ *is closed and bounded if and only if every open cover of* X *(with the induced topology) has a finite subcover.*

Motivated by this result we make the following definition.

(3.2) Definition. *A topological space* X *is compact if every open cover of* X *has a finite subcover.*

With this terminology, theorem (3.1) can be restated as follows. *The closed bounded subsets of a euclidean space are precisely those subsets which (when given the induced topology) are compact.*

The proof of theorem (3.1) will occupy us in one way or another for the next three sections. At the same time we shall build up a useful body of results on compact spaces. These spaces have some very nice properties; we state two of them now, though their proofs will have to wait until later sections:

(a) A continuous real-valued function defined on a compact space is bounded and attains its bounds.

(b) An infinite set of points in a compact space must have a limit point.

We close this section by noting that by its very definition compactness is a *topological property* of a space. That is to say, if X is compact and if X is homeomorphic to Y, then Y will be compact.

3.2 The Heine–Borel theorem

In this section we give two proofs of the celebrated Heine–Borel theorem. We include two proofs because both are interesting (the techniques involved are completely different from one another), and because the theorem lies at the heart of theorem (3.1).

(3.3) The Heine–Borel theorem. *A closed interval of the real line is compact.*

'Creeping along' proof of theorem (3.3). Let [a, b] be a closed interval of the real line, with the induced topology, and let $\mathscr{F}$ be an open cover of $[a,b]$. The idea is to 'creep along' the interval from a towards b and see how far we can get without violating the condition that our path be contained in the union of a finite number of members of $\mathscr{F}$. The theorem says that we can get all the way to b.

We define a subset X of $[a,b]$ by

$$X = \{x \in [a,b] \mid [a,x] \text{ is contained in the union of a finite subfamily of } \mathscr{F}\}.$$

Then X is nonempty ($a \in X$) and is bounded above (by b). So X has a supremum or least upper bound, say s. We claim that $s \in X$† and that $s = b$. For let O be the member of $\mathscr{F}$ which contains s. Since O is open we can choose $\varepsilon > 0$ small enough that $(s - \varepsilon, s] \subseteq O$, and if s is less than b we can assume $(s - \varepsilon, s + \varepsilon) \subseteq O$.

---

† This needs proof: the supremum of a set of real numbers need not lie in the set.

Now s is the *least* upper bound of X, consequently there are points of X arbitrarily close to s. Also, X has the property that if $x \in X$ and if $a \leqslant y \leqslant x$ then $y \in X$. Therefore we may assume $s - \varepsilon/2 \in X$. By the definition of X, the interval $[a, s - \varepsilon/2]$ is contained in the union of some finite subfamily $\mathscr{F}'$ of $\mathscr{F}$. Adding O to $\mathscr{F}'$ we obtain a finite collection of members of $\mathscr{F}$ whose union certainly contains $[a,s]$. Therefore $s \in X$. If s is less than b then $\bigcup \mathscr{F}' \cup O$ contains $[a,s + \varepsilon/2]$, giving $s + \varepsilon/2 \in X$ and contradicting the fact that s is an upper bound for X. Therefore $s = b$ and all of $[a,b]$ is contained in $\bigcup \mathscr{F}' \cup O$. This completes the proof.

'Subdivision' proof of theorem (3.3). Our second proof is less direct: we shall argue by contradiction. However, it is also less 'one-dimensional'. The same idea can be used to show, for example, that a square in the plane is a compact space.

Suppose then that theorem (3.3) is false. Let $\mathscr{F}$ be an open cover of $[a,b]$ which does not contain a finite subcover. Set $I_1 = [a,b]$. Subdivide $[a,b]$ into two closed subintervals of equal length $[a, \frac{1}{2}(a + b)]$ and $[\frac{1}{2}(a + b), b]$. At least one of these must have the property that it is not contained in the union of any finite subfamily of $\mathscr{F}$.† Select one of $[a, \frac{1}{2}(a + b)]$, $[\frac{1}{2}(a + b), b]$ which has this property and call it I_2. Now repeat the process, bisecting I_2 and selecting one half, called I_3, which is not contained in the union of any finite subfamily of $\mathscr{F}$. Continuing in this way we obtain a nested sequence of closed intervals

$$I_1 \supseteq I_2 \supseteq I_3 \supseteq \dots$$

whose lengths tend to zero as we proceed along the sequence.

We claim that $\bigcap_{n=1}^{\infty} I_n$ consists of precisely one point. In our first proof of theorem (3.3.) we used the so-called completeness property of the real numbers (in the form that a nonempty set of real numbers which is bounded above has a least upper bound) and it is at this point that we use it here. To show that the intersection of our intervals is nonempty we let x_n denote the left-hand end point of the interval I_n and we consider the sequence $\{x_n\}$. This sequence is monotonic increasing and bounded above. Therefore if p denotes the supremum of the x_n we know that $\{x_n\}$ converges to p. It is now elementary to check that $p \in I_n$ for all n. Also, since the lengths of the I_n tend to zero as n tends to infinity, it should be clear that $\bigcap_{n=1}^{\infty} I_n$ cannot contain more than one point. (The reader should make sure that he can supply the details for these statements.) Therefore $\bigcap_{n=1}^{\infty} I_n = \{p\}$.

Now p belongs to $[a,b]$ and so lies in some open set O of $\mathscr{F}$. We choose $\varepsilon > 0$ small enough that $(p - \varepsilon, p + \varepsilon) \cap [a,b] \subseteq O$, and we choose a positive integer n large enough that length $(I_n) < \varepsilon$. Since $p \in I_n$, we see that I_n is com-

---

† For if $[a, \frac{1}{2}(a + b)] \subseteq \bigcup \mathscr{F}_1$ and $[\frac{1}{2}(a + b), b] \subseteq \bigcup \mathscr{F}_2$ where $\mathscr{F}_1$ and $\mathscr{F}_2$ are both finite subfamilies of $\mathscr{F}$, then $\mathscr{F}_1 \cup \mathscr{F}_2$ is a finite subcover of $\mathscr{F}$, contradicting our assumption.

pletely contained in O. But I_n was selected so that it did not lie in the union of any finite subfamily of $\mathscr{F}$, and here we have I_n inside a single member of $\mathscr{F}$! This contradiction completes the argument.

As a corollary of theorem (3.3) we can prove that a continuous real-valued function defined on a closed interval is bounded. (We shall prove this result for a general compact space in Section 3.3.) Suppose $f:[a,b] \to \mathbb{R}$ is continuous. Given $x \in [a,b]$ we can find a neighbourhood $O(x)$ of x in $[a,b]$ such that $|f(x') - f(x)| < 1$ for all points $x' \in O(x)$. The family of all such $O(x)$ forms an open cover of $[a,b]$. Therefore by the Heine–Borel theorem we can find a finite subfamily, say $O(x_1),...,O(x_k)$, such that $O(x_1) \cup ... \cup O(x_k) = [a,b]$. Now if x lies in $O(x_i)$ then $|f(x)| \leq |f(x_i)| + 1$. So for any point x of $[a,b]$ we have

$$|f(x)| \leq \max\{|f(x_1)|,...,|f(x_k)|\} + 1$$

We mentioned earlier that the subdivision argument generalizes to higher dimensions. Consider for example the square

$$S = \{(x,y) \mid 0 \leq x \leq 1, \quad 0 \leq y \leq 1\}$$

with its usual topology induced from the plane. To show that S is compact entails proving that any family of open subsets of S whose union is all of S contains a finite subfamily whose union is also all of S. The idea, to assume the existence of a family $\mathscr{F}$ for which this is false and to work by contradiction, is exactly as before. In the subdivision process we subdivide S into four smaller squares by joining the midpoints of its opposite sides. We select one of the four which is not contained in the union of any finite subfamily of $\mathscr{F}$ and call it S_1. Repeating this process produces a nested sequence of squares

$$S \supseteq S_1 \supseteq S_2 \supseteq ...$$

whose diameters tend to zero as we move along the sequence. It is an interesting exercise to prove that $\bigcap_{n=1}^{\infty} S_n$ is exactly one point. Having done this the remainder of the argument follows as before. The details are left to the reader.

We shall give a different proof of the compactness of S in Section 3.4. The idea is quite simple: we shall define the product of two topological spaces and show that the product of compact spaces is compact. Since S is the product space $[0,1] \times [0,1]$, it will follow that S is compact.

Problems

1. Find an open cover of $\mathbb{E}^1$ which does not contain a finite subcover. Do the same for $[0,1)$ and $(0,1)$.

2. Let $S \supseteq S_1 \supseteq S_2 \supseteq ...$ be a nested sequence of squares in the plane whose diameters tend to zero as we proceed along the sequence. Prove that the intersection of all these squares consists of exactly one point.

3. Use the Heine–Borel theorem to show that an infinite subset of a closed interval must have a limit point.

4. Rephrase the definition of compactness in terms of closed sets.

3.3 Properties of compact spaces

We noted earlier that compactness is a topological property of a space, that is to say it is preserved by a homeomorphism. Even more, it is preserved by any onto continuous function.

(3.4) Theorem. *The continuous image of a compact space is compact.*

Proof. If $f:X \to Y$ is an onto continuous function, and if X is compact, then we must show Y compact. Let $\mathscr{F}$ be an open cover of Y. If $O \in \mathscr{F}$ then $f^{-1}(O)$ is an open subset of X by the continuity of f, and so the family

$$\mathscr{G} = \{f^{-1}(O) \mid O \in \mathscr{F}\}$$

is an open cover of X. Since X is compact, $\mathscr{G}$ contains a finite subcover, say $X = f^{-1}(O_1) \cup \ldots \cup f^{-1}(O_k)$. Now f is an onto function, therefore $f(f^{-1}(O_i)) = O_i$ for $1 \leqslant i \leqslant k$ and we have $Y = O_1 \cup O_2 \cup \ldots \cup O_k$. These open sets $O_1, O_2, \ldots, O_k$ are therefore a finite subcover of $\mathscr{F}$.

A subset of C of a topological space X is called a *compact subset of X* if C with the induced topology from X is a compact space. Remember that a subset U of C is open in the induced topology if and only if $U = V \cap C$ for some open set V of X. Therefore C is a compact subset of X if and only if every family of open subsets of X whose union contains C has a finite subfamily whose union also contains C.

(3.5) Theorem. *A closed subset of a compact space is compact.*

Proof. Let X be a compact space, C a closed subset of X, and $\mathscr{F}$ a family of open subsets of X such that $C \subseteq \bigcup \mathscr{F}$. If we add the open set $X - C$ to $\mathscr{F}$ we obtain an open cover of X. Using the compactness of X we know that this open cover has a finite subcover. Therefore we can find $O_1, O_2, \ldots, O_k \in \mathscr{F}$ such that $O_1 \cup O_2 \cup \ldots \cup O_k \cup (X - C) = X$. This gives $C \subseteq O_1 \cup O_2 \cup \ldots \cup O_k$, and the sets $O_1, \ldots, O_k$ provide the required finite subfamily of $\mathscr{F}$.

(3.6) Theorem. *If A is a compact subset of a Hausdorff space X, and if $x \in X - A$, then there exist disjoint neighbourhoods of x and A. Therefore a compact subset of a Hausdorff space is closed.*

Proof. Let z be a point of A. Since X is Hausdorff, we can find disjoint open sets U_z and V_z such that $x \in U_z$ and $z \in V_z$, We shall vary z in A and the notation is chosen to emphasize the dependence of U_z and V_z on z; remember x is a *fixed* point of $X - A$. Varying z throughout A produces a family of open sets $\{V_z \mid z \in A\}$ whose union contains A. But A is compact, so $A \subseteq V_{z_1} \cup \ldots \cup V_{z_k}$ for some finite collection of points $z_1, z_2, \ldots, z_k \in A$. Let $V = V_{z_1} \cup \ldots \cup V_{z_k}$. Since V_{z_i} is disjoint from the open neighbourhood U_{z_i} of x, V is disjoint from the intersection $U = U_{z_1} \cap \ldots \cap U_{z_k}$. The sets U, V are disjoint open neighbourhoods of x and A.

We have seen in Chapter 2 that a one–one onto continuous function need not have a continuous inverse, and so it need not be a homeomorphism. However, if the function goes from a compact space to a Hausdorff space then we can use the preceding results to check that its inverse is continuous.

(3.7) Theorem. *A one–one, onto, and continuous function from a compact space* X *to a Hausdorff space* Y *is a homeomorphism.*

Proof. Let $f: X \to Y$ be the function and let C be a closed subset of X. Then C is compact (theorem 3.5). Therefore $f(C)$ is compact (theorem 3.4) and consequently closed in Y (theorem 3.6). So f takes closed sets to closed sets, which proves that f^{-1} is continuous.

Our next result gives us a good feeling for the type of spaces that can be compact. It says that if we have an infinite number of points in a compact space, then the points must crowd together somewhere; in more formal language they must have a limit point.

(3.8) Bolzano–Weierstrass property. *An infinite subset of a compact space must have a limit point.*

Proof. Let X be a compact space and let S be a subset of X which has no limit point. We shall show that S is finite. Given $x \in X$ we can find an open neighbourhood $O(x)$ of x such that

$$O(x) \cap S = \begin{cases} \phi \text{ if } x \notin S \\ \{x\} \text{ if } x \in S, \end{cases}$$

since otherwise x would be a limit point of S. By the compactness of X the open cover $\{O(x) \mid x \in X\}$ has a finite subcover. But each set $O(x)$ contains at most one point of S and therefore S must be finite.

The Bolzano–Weierstrass propery tells us, for example, that a compact subset of a euclidean space cannot stretch off to infinity in some direction. For if it did, we could find infinitely many points, all well spaced out from one another and running off to infinity, with no limit point. We can of course give a precise proof of this fact using open covers of the set in question.

(3.9) Theorem. *A compact subset of a euclidean space is closed and bounded.*

Proof. Let C be a compact subset of $\mathbb{E}^n$. Then C is a closed set by theorem 3.6. Now the open balls, centre the origin with integer radius, fill out all of $\mathbb{E}^n$. Therefore if C is compact it must be contained inside the union of finitely many of these balls, i.e., there is an integer n such that C is contained in the ball with centre the origin and radius n. In other words C is bounded.

(3.10) Theorem. *A continuous real-valued function defined on a compact space is bounded and attains its bounds.*

Proof. If $f: X \to \mathbb{R}$ is continuous and if X is compact, then $f(X)$ is compact. Therefore $f(X)$ is a closed bounded subset of $\mathbb{R}$ by theorem (3.9) and f is certainly bounded. Since $f(X)$ is closed, both the supremum and infimum of $f(X)$ lie in $f(X)$. We can therefore find points $x_1, x_2 \in X$ such that

$$f(x_1) = \sup(f(X)) \quad \text{and} \quad f(x_2) = \inf(f(X)),$$

which says precisely that f attains its bounds.

We end this section with a rather technical result concerning open covers of a compact metric space: the result will be applied several times in later chapters.

(3.11) Lebesgue's lemma. *Let X be a compact metric space and let $\mathscr{F}$ be an open cover of X. Then there exists a real number $\delta > 0$ (called a Lebesgue number of $\mathscr{F}$) such that any subset of X of diameter less than δ is contained in some member of $\mathscr{F}$.*

Proof. If Lebesgue's lemma is false we can find a sequence $A_1, A_2, A_3, \ldots$ of subsets of X, none of which are contained inside a member of $\mathscr{F}$, and whose diameters tend to zero as we proceed along the sequence. For each n choose a point x_n belonging to A_n. Either the sequence $\{x_n\}$ contains only finitely many distinct points, in which case some point repeats infinitely often; or it is infinite, in which case it must have a limit point since X is compact. Denote the repeated point, or limit point, by p. Let U be an element of $\mathscr{F}$ which contains p. Choose $\varepsilon > 0$ such that $B(p,\varepsilon) \subseteq U$, and choose an integer N large enough so that:
(a) the diameter of A_N is less than $\varepsilon/2$, and
(b) $x_N \in B(p, \varepsilon/2)$.
Then $d(x_N, p) < \varepsilon/2$ and $d(x, x_N) < \varepsilon/2$ for any point x of A_N. Therefore $d(x,p) < \varepsilon$ if $x \in A_N$, showing $A_N \subseteq U$. This contradicts our initial choice of the sequence $\{A_n\}$.

Problems

5. Which of the following are compact? (a) the space of rational numbers; (b) S^n with a finite number of points removed; (c) the torus with an open disc

removed; (d) the Klein bottle; (e) the Möbius strip with its boundary circle removed.

6. Show that the Hausdorff condition cannot be relaxed in theorem (3.7).

7. Show that Lebesgue's lemma fails for the plane.

8. (*Lindelöf's theorem*). If X has a countable base for its topology, prove that any open cover of X contains a countable subcover.

9. Prove that two disjoint compact subsets of a Hausdorff space always possess disjoint neighbourhoods.

10. Let A be a compact subset of a metric space X. Show that the diameter of A is equal to $d(x,y)$ for some pair of points $x,y \in A$. Given $x \in X$, show that $d(x,A) = d(x,y)$ for some $y \in A$. Given a closed subset B, disjoint from A, show that $d(A,B) > 0$.

11. Find a topological space and a compact subset whose closure is not compact.

12. Do the real numbers with the finite-complement topology form a compact space? Answer the same question for the *half-open interval topology* (see Problem 11 of Chapter 2).

13. Let $f:X \to Y$ be a closed map with the property that the inverse image of each point of Y is a compact subset of X. Show that $f^{-1}(K)$ is compact whenever K is compact in Y. Can you remove the condition that f be closed?

14. If $f:X \to Y$ is a one–one map, and if $f:X \to f(X)$ is a homeomorphism when we give $f(X)$ the induced topology from Y, we call f an *embedding* of X in Y. Show that a one–one map from a compact space to a Hausdorff space must be an embedding.

15. A space is *locally compact* if each of its points has a compact neighbourhood. Show that the following are all locally compact: any compact space; $\mathbb{E}^n$; any discrete space; any closed subset of a locally compact space. Show that the space of rationals is not locally compact. Check that local compactness is preserved by a homeomorphism.

16. Suppose X is locally compact and Hausdorff. Given $x \in X$ and a neighbourhood U of x, find a compact neighbourhood of x which is contained in U.

17. Let X be a locally compact Hausdorff space which is not compact. Form a new space by adding one extra point, usually denoted by ∞, to X and taking the open sets of $X \cup \{\infty\}$ to be those of X together with sets of the form $(X - K) \cup \{\infty\}$, where K is a compact subset of X. Check the axioms for a topology, and show that $X \cup \{\infty\}$ is a compact Hausdorff space which contains X as a dense subset. The space $X \cup \{\infty\}$ is called the *one-point compactification* of X.

18. Prove that $\mathbb{E}^n \cup \{\infty\}$ is homeomorphic to S^n. (Think first of the case $n = 2$. Stereographic projection gives a homeomorphism between $\mathbb{E}^2$ and S^2 minus the north pole, points 'out towards infinity' in the plane becoming points near to the north pole on the sphere. Think of replacing the north pole in S^2 as adding a point at ∞ to $\mathbb{E}^2$.)

19. Let X and Y be locally compact Hausdorff spaces and let $f: X \to Y$ be an onto map. Show f extends to a map from $X \cup \{\infty\}$ onto $Y \cup \{\infty\}$ iff $f^{-1}(K)$ is compact for each compact subset K of Y. Deduce that if X and Y are homeomorphic spaces then so are their one-point compactifications. Find two spaces which are not homeomorphic but which have homeomorphic one-point compactifications.

3.4 Product spaces

We now turn to the study of spaces which have a natural *product* structure. Examples spring readily to mind: we can think of the plane as the product of two copies of the real line, the torus as the product of two circles, or the cylinder as the product of a circle with the unit interval. It is worth looking at one of these examples in more detail. Take a specific cylinder in $\mathbb{E}^3$, say

$$\{(x,y,z) \mid x^2 + y^2 = 1 \quad \text{and} \quad 0 \leqslant z \leqslant 1\}$$

and give it the induced topology. As a *set* it is the cartesian product $S^1 \times I$, where S^1 denotes the unit circle in the (x,y) plane and I the unit interval on the z axis. We claim that the topology of the cylinder is, in a very natural sense, the product of the topologies of the circle and the interval. To see this, we note that if U is an open set in S^1, and if V is open in I, then the product $U \times V$ is open in the cylinder (Fig. 3.1). Also, if we are given an open set O of the cylinder, and a point

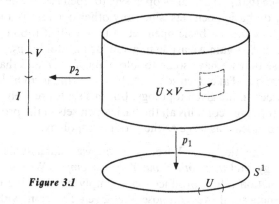

Figure 3.1

p belonging to O, then we can easily find open sets $U \subseteq S^1$, $V \subseteq I$ such that $p \in U \times V \subseteq O$. In other words, these product sets $U \times V$ form a base for the topology of the cylinder. We summarize this by saying that the cylinder has the 'product topology'. Motivated by the above, we can give a precise definition of the product of two topological spaces. Once this is done we shall prove (as the main result of this section) that the product of two compact spaces is compact.

Let X and Y be topological spaces and let $\mathscr{B}$ denote the family of all subsets of $X \times Y$ of the form $U \times V$, where U is open in X and V open in Y. Then

51

$\bigcup \mathscr{B} = X \times Y$ and the intersection of any two members of $\mathscr{B}$ lies in $\mathscr{B}$. Therefore $\mathscr{B}$ is a base for a topology on $X \times Y$. This topology is called the *product topology* and the set $X \times Y$, when equipped with the product topology, is called a *product space*. We need hardly mention that the same construction goes through for a finite product. If $X_1, X_2, \ldots, X_n$ are topological spaces, the product topology on $X_1 \times X_2 \times \ldots \times X_n$ has as base the sets $U_1 \times U_2 \times \ldots \times U_n$, where U_i is open in X_i, $1 \leqslant i \leqslant n$. We note that the natural topology of euclidean n-space is precisely the product topology relative to the decomposition of $\mathbb{E}^n$ as the product of n copies of the real line. For simplicity, we shall work with the product of two spaces, but we emphasize that all the results (and proofs!) go through for finite products. In fact, since $X_1 \times X_2 \times \ldots \times X_n$ is clearly homeomorphic to $(X_1 \times X_2 \times \ldots \times X_{n-1}) \times X_n$, results for finite products follow by induction from those for the product of two spaces.

The functions $p_1 : X \times Y \to X$ and $p_2 : X \times Y \to Y$ defined by $p_1(x,y) = x$, $p_2(x,y) = y$ are called *projections*. We can characterize the product topology in terms of these projections as follows.

(3.12) Theorem. *If* $X \times Y$ *has the product topology then the projections are continuous functions and they take open sets to open sets. The product topology is the smallest topology on* $X \times Y$ *for which both projections are continuous.*

Proof. Suppose U is an open subset of X, then $p_1^{-1}(U) = U \times Y$, which is open in the product topology: therefore p_1 is continuous. The argument for p_2 is similar. To see that p_1, say, takes open sets to open sets we need only look at the effect of p_1 on basic open sets, since any other open set is a union of these. But $p_1(U \times V) = U$, so a basic open set of the product topology is sent by p to an open set in X. Again we argue in a similar fashion for p_2.

Now suppose that we have some topology on $X \times Y$ and that both projections are continuous. Take open sets $U \subseteq X$, $V \subseteq Y$ and form $p_1^{-1}(U) \cap p_2^{-1}(V)$. This must be open in the given topology. But this set is precisely $U \times V$. Therefore the given topology contains all the basic open sets of the product topology, and is therefore at least as large as the product topology.

Whenever we mention $X \times Y$ from now on, we shall assume that it has the product topology, and *that both X and Y are nonempty*. We can check the continuity of a function into a product space simply by checking that we obtain continuous functions if we compose the given function with each of the projections.

(3.13) Theorem. *A function* $f : Z \to X \times Y$ *is continuous if and only if the two composite functions* $p_1 f : Z \to X$, $p_2 f : Z \to Y$ *are both continuous.*

Proof. Suppose that both $p_1 f$ and $p_2 f$ are continuous. To check the continuity of f we need only show that $f^{-1}(U \times V)$ is open in Z for each basic open set $U \times V$ of $X \times Y$. But

$$f^{-1}(U \times V) = (p_1 f)^{-1}(U) \cap (p_2 f)^{-1}(V)$$

the intersection of two open subsets of Z. Therefore $f^{-1}(U \times V)$ is open in Z. Conversely, if f is continuous then $p_1 f$ and $p_2 f$ are continuous, by the continuity of the projections p_1, p_2.

(3.14) Theorem. *The product space* X × Y *is a Hausdorff space if and only if both* X *and* Y *are Hausdorff.*

Proof. Suppose that X and Y are both Hausdorff spaces. Let (x_1, y_1) and (x_2, y_2) be distinct points of $X \times Y$. Then either $x_1 \neq x_2$ or $y_1 \neq y_2$ (or both): assume for the sake of argument that $x_1 \neq x_2$. Since X is Hausdorff we can find disjoint open sets U_1, U_2 in X such that $x_1 \in U_1$ and $x_2 \in U_2$. To find disjoint open neighbourhoods of (x_1, y_1) and (x_2, y_2) we simply form the products $U_1 \times Y$, $U_2 \times Y$.

Conversely, suppose that $X \times Y$ is Hausdorff. Given distinct points $x_1, x_2 \in X$, we choose a point $y \in Y$ and find disjoint basic open sets $U_1 \times V_1$, $U_2 \times V_2$ in $X \times Y$ such that $(x_1, y) \in U_1 \times V_1$ and $(x_2, y) \in U_2 \times V_2$. Then U_1, U_2 are disjoint open neighbourhoods of x_1 and x_2 in X. Therefore X is a Hausdorff space. The argument for Y is similar.

(3.15) Theorem. X × Y *is compact if and only if both* X *and* Y *are compact.*

(3.16) Lemma. *Let* X *be a topological space and let* $\mathscr{B}$ *be a base for the topology of* X. *Then* X *is compact if and only if every open cover of* X *by members of* $\mathscr{B}$ *has a finite subcover.*

Proof of the lemma. Suppose that every open cover of X by members of $\mathscr{B}$ has a finite subcover, and let $\mathscr{F}$ be an arbitrary open cover of X. Since $\mathscr{B}$ is a base for the topology of X we know that we can express each member of $\mathscr{F}$ as a union of members of $\mathscr{B}$. Let $\mathscr{B}'$ denote the family of those members of $\mathscr{B}$ which are used in this process. By construction we have $\bigcup \mathscr{B}' = \bigcup \mathscr{F} = X$; so $\mathscr{B}'$ is an open cover of X (by members of $\mathscr{B}$) and must therefore contain a finite subcover. For each basic open set in this finite subcover, we select a single member of $\mathscr{F}$ which contains it. This gives a finite subcover of $\mathscr{F}$ and shows that X is compact. The converse is obvious.

Proof of theorem (3.15). If $X \times Y$ is compact, then both X and Y have to be compact since the projections $p_1 : X \times Y \to X$, $p_2 : X \times Y \to Y$ are onto and continuous functions. (Remember we have assumed both X and Y are non-empty.)

Now for the more interesting part of the result: suppose both X and Y are compact spaces and let $\mathscr{F}$ be an open cover of $X \times Y$ by *basic* open sets of the form $U \times V$, where U is open in X and V open in Y. We shall show that $\mathscr{F}$ must contain a finite subcover. This is enough to show $X \times Y$ compact by the previous lemma.

Select a point $x \in X$ and consider the subset $\{x\} \times Y$ of $X \times Y$ with the induced topology. It is easy to check that

$$p_2|\{x\} \times Y : \{x\} \times Y \to Y$$

is a homeomorphism. In other words $\{x\} \times Y$ is just a copy of Y in our product space which lies 'over' the point x (see Fig. 3.2). So $\{x\} \times Y$ is compact and we can find a minimal finite subfamily of $\mathscr{F}$ whose union contains $\{x\} \times Y$. We shall label the members of this finite subfamily

$$U_1^x \times V_1^x, U_2^x \times V_2^x, \dots, U_{n_x}^x \times V_{n_x}^x$$

in order to emphasize their dependence on the point x. Note that the union of these sets contains more than $\{x\} \times Y$, it actually contains all of $U^x \times Y$ where $U^x = \bigcap_{i=1}^{n_x} U_i^x$.

So far we have only made use of the compactness of Y. Now the set $U^x \times Y$ appears in Fig. 3.2 as a strip in $X \times Y$ lying over the subset U^x of X. The idea

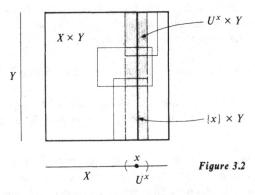

Figure 3.2

of the remainder of the proof is to use the fact that X is compact to show that we can cover all of $X \times Y$ by a finite number of such strips. The family $\{U^x \mid x \in X\}$ is an open cover of X and we select from it a finite subcover, say

$$U^{x_1}, U^{x_2}, \dots, U^{x_s}$$

Since X is the union of these sets we have

$$X \times Y = (U^{x_1} \times Y) \cup (U^{x_2} \times Y) \cup \dots \cup (U^{x_s} \times Y)$$

But $U^{x_i} \times Y$ is contained in $(U_1^{x_i} \times V_1^{x_i}) \cup \dots \cup (U_{n_{x_i}}^{x_i} \times V_{n_{x_i}}^{x_i})$. Therefore the basic open sets

$$U_1^{x_i} \times V_1^{x_i}, U_2^{x_i} \times V_2^{x_i}, \dots, U_{n_{x_i}}^{x_i} \times V_{n_{x_i}}^{x_i}, \qquad 1 \leqslant i \leqslant s$$

form a finite subcover of $\mathscr{F}$. This completes the proof.

We are now in a position to prove theorem (3.1) and complete our characterization of closed bounded subsets of a euclidean space. We recall the statement of the theorem:

(3.1) *Theorem. A subset of* $\mathbb{E}^n$ *is compact if and only if it is closed and bounded.*

Proof. We have already shown, in theorem (3.9), that a compact subset of a euclidean space is both closed and bounded. Suppose, conversely, that X is a closed bounded subset of $\mathbb{E}^n$. We think of $\mathbb{E}^n$ as the product of n copies of the real line, and note that since X is bounded it must be contained in

$$[-s,s] \times [-s,s] \times \ldots \times [-s,s]$$

(the product of n copies of the closed interval $[-s,s]$) for some real number s. The Heine–Borel theorem tells us that $[-s,s]$ is compact and theorem (3.15) shows the product of any finite number of copies of this interval to be compact. Therefore X is a closed subset of a compact space, and hence compact by theorem (3.5).

Before leaving the notion of compactness, we should mention that it is possible to define the product of an infinite collection of topological spaces, and to prove that any product of compact spaces is compact. This result is usually referred to as the Tychonoff theorem; it is considerably deeper than the finite version, theorem (3.15), being equivalent to the Axiom of Choice. For details of the Tychonoff theorem we refer the reader to Kelley [17].

Problems

20. If $X \times Y$ has the product topology, and if $A \subseteq X$, $B \subseteq Y$, show that $\overline{A \times B} = \bar{A} \times \bar{B}$, $(A \times B)^\circ = \mathring{A} \times \mathring{B}$, and $\mathrm{Fr}(A \times B) = [\mathrm{Fr}(A) \times \bar{B}] \cup [\bar{A} \times \mathrm{Fr}(B)]$ where $\mathrm{Fr}(\ )$ denotes frontier.

21. If A and B are compact, and if W is a neighbourhood of $A \times B$ in $X \times Y$, find a neighbourhood U of A in X and a neighbourhood V of B in Y such that
$$U \times V \subseteq W.$$

22. Prove that the product of two second-countable spaces is second-countable, and that the product of two separable spaces is separable.

23. Prove that $[0,1) \times [0,1)$ is homeomorphic to $[0,1] \times [0,1)$.

24. Let $x_0 \in X$ and $y_0 \in Y$. Prove that the functions $f:X \to X \times Y$, $g:Y \to X \times Y$ defined by $f(x) = (x,y_0)$, $g(y) = (x_0,y)$ are embeddings (as defined in Problem 14).

25. Show that the diagonal map $\Delta:X \to X \times X$ defined by $\Delta(x) = (x,x)$ is indeed a map, and check that X is Hausdorff iff $\Delta(X)$ is closed in $X \times X$.

26. We know that the projections $p_1:X \times Y \to X$, $p_2:X \times Y \to Y$ are open maps. Are they always closed?

27. Given a countable number of spaces $X_1, X_2, \ldots$, a typical point of the product ΠX_i will be written $x = (x_1, x_2, \ldots)$. The *product topology* on ΠX_i is the smallest topology for which all of the projections $p_i:\Pi X_i \to X_i$, $p_i(x) = x_i$, are continuous. Construct a base for this topology from the open sets of the spaces $X_1, X_2, \ldots$.

28. If each X_i is a metric space, the topology on X_i being induced by a metric d_i, prove that

$$d(x,y) = \sum_{i=1}^{\infty} \frac{1}{2^i} \frac{d_i(x_i, y_i)}{1 + d_i(x_i, y_i)}$$

defines a metric on ΠX_i which induces the product topology.

29. The *box topology* on ΠX_i has as base all sets of the form $U_1 \times U_2 \times \ldots$, where U_i is open in X_i. Show that the box topology contains the product topology, and that the two are equal iff X_i is an indiscrete space for all but finitely many values of i. (X is an *indiscrete space* if the only open sets are $\varnothing$ and X.)

3.5   Connectedness

A space such as the real line, or the torus, seems to be connected, i.e., to be all in one piece. It is not hard to give a precise definition of this intuitive idea of connectedness and to see that it is a topological property of a space.

We have already said that being connected means, intuitively, being all in one piece. So if X is a connected space, and if we write X as the union $A \cup B$ of two nonempty subsets, then we expect A and B either to intersect or at the very least to abut against one another in X. We can express this mathematically by asking that one of

$$\bar{A} \cap B, \qquad A \cap \bar{B}$$

be nonempty: in other words, either A and B have a point in common, or some point of B is a limit point of A, or some point of A is a limit point of B. For example, if we decompose the closed interval $[0,1]$ as $[0,\frac{1}{2}) \cup [\frac{1}{2},1]$ then the point $\frac{1}{2}$ lies in $\overline{[0,\frac{1}{2})} \cap [\frac{1}{2},1]$.

(3.17) Definition. *A space* X *is connected if whenever it is decomposed as the union* A $\cup$ B *of two nonempty subsets then* $\bar{A} \cap B \neq \varnothing$ *or* A $\cap \bar{B} \neq \varnothing$.

(3.18) Theorem. *The real line is a connected space.*

Proof. Suppose $\mathbb{R} = A \cup B$, where both A and B are nonempty and $A \cap B = \varnothing$. We shall show that some point of A is a limit point of B, or that some point of B is a limit point of A. Choose points $a \in A$, $b \in B$, and (without loss of generality) suppose that $a < b$. Let X consist of those points of A which are less than b and let s denote the supremum of X. This point s may or may not lie in A; however, if s does not lie in A then, by the very definition of supremum, s must lie in $\bar{A}$. We shall consider these two possibilities separately. Suppose s lies in

A, then $s < b$, and since s is an upper bound for X, all the points between s and b lie in B. Therefore s is a limit point of B. If s does not belong to A, then automatically s lies in B as A and B fill out all of $\mathbb{R}$. We noted above that in this case s is a limit point of A. Therefore we have shown that either $\bar{A}$ intersects B or A intersects $\bar{B}$.

As usual, we say that a subset of a topological space is connected if it becomes a connected space when given the induced topology. We shall call a subset X of the real line an *interval* if, whenever we have distinct points $a, b \in X$, then all points which are greater than a and less than b also lie in X. This is the usual notion of interval: it includes the possibility that an interval be open, closed, half open, or that it stretch off to infinity in some direction. Our intuition suggests very strongly that the intervals should be the only connected subsets of the real line. All other subsets have 'gaps' in them, and therefore consist of several distinct pieces.

(3.19) Theorem. *A nonempty subset of the real line is connected if and only if it is an interval.*

Proof. The proof of theorem (3.18) adapts very easily to show that any interval is connected. If X is not an interval, then we can find points $a, b \in X$ and a point p which lies outside X yet nevertheless satisfies $a < p < b$. Let A denote the subset of X consisting of those points which are less than p, and let $B = X - A$. Since p is not in X, every point of the closure of A in X is less than p, and every point of the closure of B in X is greater than p. Therefore $\bar{A} \cap B$ and $A \cap \bar{B}$ are both empty and we see that X is not connected.

The definition of connectedness can be formulated in more than one way.

(3.20) Theorem. *The following conditions on a space* X *are equivalent:*

(a) X *is connected.*
(b) *The only subsets of* X *which are both open and closed are* X *and the empty set.*
(c) X *cannot be expressed as the union of two disjoint nonempty open sets.*
(d) *There is no onto continuous function from* X *to a discrete space which contains more than one point.*

Proof. We shall show that (a) $\Rightarrow$ (b) $\Rightarrow$ (c) $\Rightarrow$ (d) $\Rightarrow$ (a). Suppose X is connected and let A be a subset of X which is both open and closed. If $B = X - A$ then B is also both open and closed. Since both A and B are closed we have $\bar{A} = A$ and $\bar{B} = B$, giving $\bar{A} \cap B = A \cap \bar{B} = A \cap B = \varnothing$. But X is connected, so one of A, B must be empty and the other one the whole space. This proves (a) $\Rightarrow$ (b). The implication (b) $\Rightarrow$ (c) is obvious.

Now suppose (c) is satisfied, and let Y be a discrete space with more than one point and let $f : X \to Y$ be an onto continuous function. Break up Y as a union $U \cup V$ of two disjoint nonempty open sets. Then $X = (f^{-1}U) \cup (f^{-1}V)$, contradicting (c).

We are left to show (d) $\Rightarrow$ (a). Let X be a space which satisfies (d) and suppose X is not connected. Decompose X as $A \cup B$ where A and B are nonempty and satisfy $\bar{A} \cap B = A \cap \bar{B} = \varnothing$. We notice that both A and B are open sets, for example, B is the complement of the closed set $\bar{A}$, and we define a function f from X to the subspace $\{-1,1\}$ of the real line by

$$f(x) = \begin{cases} -1 & \text{if} \quad x \in A \\ 1 & \text{if} \quad x \in B. \end{cases}$$

Then f is continuous and onto, contradicting (d) for X.

A continuous function should not be able to tear a space into pieces (i.e., send a connected space onto a space which is not connected): we expect quite the reverse, namely that a continuous function should preserve connectedness.

(3.21) Theorem. *The continuous image of a connected space is connected.*

Proof. Let $f : X \to Y$ be an onto continuous function and suppose that X is connected. If A is a subset of Y which is both open and closed, then $f^{-1}(A)$ is open and closed in X. Since X is connected $f^{-1}(A)$ must be all of X or the empty set, by condition (b) of theorem (3.20). Therefore A is equal to Y or empty and we have proved Y to be connected.

(3.22) Corollary. *If* $h : X \to Y$ *is a homeomorphism, then* X *is connected if and only if* Y *is connected. In brief, connectedness is a topological property of a space.*

(3.23) Theorem. *Let* X *be a topological space and let* Z *be a subset of* X. *If* Z *is connected, and if* Z *is dense in* X, *then* X *is connected.*

Proof. Let A be a nonempty subset of X which is both open and closed. Since Z is dense in X we know that Z must intersect every nonempty open subset of X, and therefore $A \cap Z$ is nonempty. Now $A \cap Z$ is both open and closed in Z, and since Z is connected we deduce that $A \cap Z = Z$, i.e., $Z \subseteq A$. Therefore $X = \bar{Z} \subseteq \bar{A} = A$, giving $X = A$ as required.

(3.24) Corollary. *If* Z *is a connected subset of a topological space* X, *and if* $Z \subseteq Y \subseteq \bar{Z}$, *then* Y *is connected. In particular, the closure* $\bar{Z}$ *of* Z *is connected.*

Proof. Notice that the closure of Z in Y is all of Y and apply theorem (3.23) to the pair $Z \subseteq Y$.

We need a little more terminology. If A and B are subsets of a space X, and if $\bar{A} \cap \bar{B}$ is empty, we say that A and B are *separated* from one another in X.

(3.25) Theorem. *Let* $\mathscr{F}$ *be a family of subsets of a space* X *whose union is all of* X. *If each member of* $\mathscr{F}$ *is connected, and if no two members of* $\mathscr{F}$ *are separated from one another in* X, *then* X *is connected.*

Proof. Let A be a subset of X which is both open and closed. We shall show that A is either empty or equal to all of X. Each member of $\mathscr{F}$ is connected, so if $Z \in \mathscr{F}$ we know that $Z \cap A$ is either empty or all of Z. If $Z \cap A = \varnothing$ for all Z in $\mathscr{F}$ then $A = \varnothing$. The other possibility is that we can find some element $Z \in \mathscr{F}$ for which $Z \cap A = Z$, i.e., for which Z is contained in A. Suppose W is some other element of $\mathscr{F}$. If $W \cap A$ is empty, then W and Z are separated from one another in X. (For $W \cap A = \varnothing$ gives $\bar{W} \subseteq \overline{X - A}$ and since $X - A$ is closed we have $\bar{W} \subseteq X - A$. Now combine this with $\bar{Z} \subseteq \bar{A} = A$.) However, we are told that no two subsets of $\mathscr{F}$ are separated from one another in X. Therefore $W \subseteq A$ for all $W \in \mathscr{F}$ and $A = \bigcup \mathscr{F} = X$.

(3.26) Theorem. *If* X *and* Y *are connected spaces then the product space* $X \times Y$ *is connected.*

Proof. If x is a point of X, the subspace $\{x\} \times Y$ of $X \times Y$ is connected since it is homeomorphic to Y. Similarly $X \times \{y\}$ is connected for any point y of Y. Now $\{x\} \times Y$ and $X \times \{y\}$ overlap in the point (x,y), therefore $Z(x,y) = (\{x\} \times Y) \cup (X \times \{y\})$ is connected. (Apply theorem (3.25) to the space $Z(x,y)$.) Also $X \times Y = \bigcup_{\substack{x \in X \\ y \in Y}} Z(x,y)$, and any two of the $Z(x,y)$ have nonempty intersection. Therefore a second application of theorem (3.25) shows $X \times Y$ to be connected. Figure 3.3 illustrates this proof.

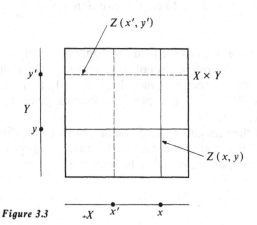

Figure 3.3

This last result tells us immediately that euclidean n-space is connected, since it is the product of a finite number of copies of the real line. Now consider the unit sphere S^n in $\mathbb{E}^{n+1}$ where $n \geqslant 1$. If we remove a point from S^n we obtain a space homeomorphic to $\mathbb{E}^n$. But the closure of S^n minus a point is all of S^n when $n \geqslant 1$. Therefore S^n is a connected space for $n \geqslant 1$, by theorem (3.23). We also see that the torus is connected, since we can think of it as the product space $S^1 \times S^1$.

We should point out that if a product space $X \times Y$ is connected, and if $X \times Y$ is nonempty, then the factors X and Y have to be connected. This follows from the continuity of the projections.

If a space is not connected then it breaks up as a union of connected pieces, any two of which are separated from one another. We call these pieces *components*. More formally, a component of a topological space X is a maximal connected subset of X.

(3.27) Theorem. *Each component of a topological space is a closed set and distinct components are separated from one another in the space.*

Proof. Let C be a component of X. Then C is connected, and so $\bar{C}$ is connected by corollary (3.24). But C is a maximal connected subset of X, therefore $C = \bar{C}$ and we see that C is closed. If D is some other component of X, and if D is not separated from C in X, then $C \cup D$ is connected by theorem (3.25). This contradicts the maximality of C (and D).

We note that every connected subset of a space is contained in a component. For if $A \subseteq X$ and if A is connected, then define C to be the union of the family of all connected subsets of X which contain A. This set C is connected by theorem (3.25) and is maximal by its very construction. Therefore C is a component which contains A.

One or two examples should help the intuition along.

Examples.
1. A connected space, such as the torus, has only one component. At the other extreme, each point of a discrete topological space is a component of the space.
2. $\mathbb{E}^1 - S^0$ has three components, namely $(-\infty, -1), (-1, 1)$, and $(1, \infty)$. For $n \geqslant 1$ the space $\mathbb{E}^{n+1} - S^n$ has two components given by the conditions $\|x\| > 1$ and $\|x\| < 1$.
3. Each point of the rationals $\mathbb{Q}$ (with the induced topology from the real line) is a component. Note that $\mathbb{Q}$ is not a discrete space. A space like this, in which every point is a component, is said to be *totally disconnected*.

Problems

30. Let X be the set of all points in the plane which have at least one rational coordinate. Show that X, with the induced topology, is a connected space.

31. Give the set of real numbers the finite-complement topology. What are the components of the resulting space? Answer the same question for the half-open interval topology.

32. If X has only a finite number of components, show that each component is both open and closed. Find a space none of whose components are open sets.

33. (*Intermediate value theorem*). If $f:[a,b] \to \mathbb{E}^1$ is a map such that $f(a) < 0$ and $f(b) > 0$, use the connectedness of $[a,b]$ to establish the existence of a point c for which $f(c) = 0$.

34. A space X is *locally connected* if for each $x \in X$, and each neighbourhood U of x, there is a connected neighbourhood V of x which is contained in U. Show that any euclidean space, and therefore any space which is locally euclidean (like a surface), is locally connected. If $X = \{0\} \cup \{1/n \mid n = 1,2,...\}$ with the subspace topology from the real line, show that X is not locally connected.

35. Show that local connectedness is preserved by a homeomorphism, but need not be preserved by a continuous function.

36. Show that X is locally connected iff every component of each open subset of X is an open set.

3.6   Joining points by paths

A *path* in a topological space X is a continuous function $\gamma:[0,1] \to X$. The points $\gamma(0)$ and $\gamma(1)$ are called the *beginning* and *end* points of the path respectively, and γ is said to *join* $\gamma(0)$ to $\gamma(1)$. Note that if γ^{-1} is defined by $\gamma^{-1}(t) = \gamma(1 - t)$, $0 \leqslant t \leqslant 1$, then γ^{-1} is a path in X which joins $\gamma(1)$ to $\gamma(0)$.

(3.28) Definition. *A space is path-connected if any two of its points can be joined by a path.*

If γ is a path in X, and if $f:X \to Y$ is a continuous function, then the composition

$$[0,1] \xrightarrow{\gamma} X \xrightarrow{f} Y$$

is a path in Y. From this remark it should be clear that if $h:X \to Y$ is a homeomorphism, and if X is path-connected, then Y is also path-connected. In other words, the property of being path-connected is, like compactness and connectedness, a topological property of a space.

A path-connected space is always connected, but the converse is not true. We shall often require our spaces to be path-connected. This is a natural condition to impose, for example, when working with the fundamental group of a space, since the elements of the fundamental group are constructed using paths in the space.

(3.29) Theorem. *A path-connected space is connected.*

Proof. Let X be a path-connected space and let A be a nonempty subset of X which is both open and closed in X. Assume A is not all of X, choose points $x \in A$, $y \in X - A$, and join x to y by a path γ in X. Then $\gamma^{-1}(A)$ is a nonempty

61

proper subset of $[0,1]$ which (by the continuity of γ) is both open and closed. This contradicts the fact that $[0,1]$ is connected. Our assumption $A \neq X$ must therefore be false, and we have $A = X$ as required.

Note that if we have points x,y,z in a space X, and paths α,β joining x to y and y to z respectively, then the path γ defined by

$$\gamma(t) = \begin{cases} \alpha(2t) & 0 \leqslant t \leqslant \frac{1}{2} \\ \beta(2t - 1) & \frac{1}{2} \leqslant t \leqslant 1 \end{cases}$$

joins x to z.

(3.30) Theorem. *A connected open subset of a euclidean space is path-connected.*

Proof. Let X be a connected open subset of $\mathbb{E}^n$. Given $x \in X$, we denote by $U(x)$ the collection of those points of X which can be joined to x by a path in X. Our aim will be to show that $U(x)$ is all of X. Since $U(x)$ is quite clearly path-connected, this will prove the theorem. Let $y \in U(x)$ and choose a ball B with centre y which lies entirely in X. If $z \in B$ then we can join z to x by a path in X, for we can join z to y by a straight line in B and follow this by a path from y to x. Therefore B is contained in $U(x)$ and we see that $U(x)$ is open in X. Also, the complement of $U(x)$ in X is the union of the family $\{U(y) \mid y \in X - U(x)\}$, and is therefore open. So $U(x)$ is closed in X. Since X is connected and $U(x)$ is nonempty (it contains at least the point x) we have $U(x) = X$.

We mentioned earlier the existence of spaces which are connected, yet not path-connected: Fig. 3.4 illustrates a compact subspace of the plane with these properties. Define

$$Y = \{(0,y) \in \mathbb{E}^2 \mid -1 \leqslant y \leqslant 1\}$$

$$Z = \left\{ \left(x, \sin\frac{\pi}{x}\right) \in \mathbb{E}^2 \mid 0 < x \leqslant 1 \right\}$$

and set $X = Y \cup Z$. Now Z is a connected space because it is the image of $(0,1]$ under a continuous function. It is easy to check that the closure of Z in $\mathbb{E}^2$ is precisely X, so X is connected. To show that X is not path-connected, we shall prove that it is impossible to join a point of Y to a point of Z by a path in X. Let $y \in Y$ and let $\gamma:[0,1] \to X$ be a path which begins at y. Since Y is closed in $\mathbb{E}^2$ it is a closed subset of X, and therefore $\gamma^{-1}(Y)$ is closed in $[0,1]$. Now $\gamma^{-1}(Y)$ is certainly nonempty (it contains 0), so if we can show it is open in $[0,1]$ we will have $\gamma^{-1}(Y) = [0,1]$, i.e., $\gamma([0,1]) \subseteq Y$, as required. Suppose $t \in \gamma^{-1}(Y)$ and choose $\varepsilon > 0$ small enough so as to ensure that $\gamma((t - \varepsilon, t + \varepsilon))$ is contained in the closed disc D, centre $\gamma(t)$ and radius $\frac{1}{2}$. The intersection of this disc with our space X consists of a closed interval on the y axis, together with segments of the curve $y = \sin\frac{\pi}{x}$, each of which is homeomorphic to a closed interval. Furthermore, any two of these sets are separated from one another in $D \cap X$. Therefore $D \cap Y$ is a component of $D \cap X$. Since $\gamma(t) \in D \cap Y$, and $(t - \varepsilon, t + \varepsilon)$ is connected, we must have all of $\gamma((t - \varepsilon, t + \varepsilon))$ in $D \cap Y$. This

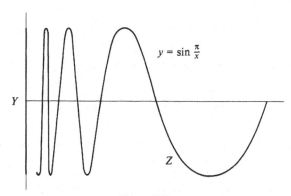

Figure 3.4

proves $\gamma^{-1}(Y)$ is open in $[0,1]$ and completes our verification that X is not path-connected.

A *path component* of a space X is (by analogy with the notion of component) a maximal path-connected subset of X. Each path component is connected and therefore lies inside a component. However, path components are not in general separated from one another, nor are they necessarily closed. For example, the path components of the space shown in Fig. 3.4 are precisely the sets Y, Z. These are not separated from one another, and Z is not closed.

Problems

37. Show that the continuous image of a path-connected space is path-connected.

38. Show that S^n is path-connected for $n > 0$.

39. Prove that the product of two path-connected spaces is path-connected.

40. If A and B are path-connected subsets of a space, and if $A \cap B$ is nonempty, prove that $A \cup B$ is path-connected.

41. Find a path-connected subset of a space whose closure is not path-connected.

42. Show that any indiscrete space is path-connected.

43. A space X is *locally path-connected* if for each $x \in X$, and each neighbourhood U of x, there is a path-connected neighbourhood V of x which is contained in U. Is the space shown in Fig. 3.4 locally path-connected? Convert the space

$\{0\} \cup \{1/n \mid n = 1,2,\ldots\}$ into a subspace of the plane which is path-connected but not locally path-connected.

44. Prove that a space which is connected and locally path-connected is path-connected.

4. Identification Spaces

4.1 Constructing a Möbius strip

Many interesting spaces can be constructed as follows. Begin with a fairly simple topological space X and produce a new space by identifying some of the points of X. We have already made use of this process: in Chapter 1 we had occasion to construct various surfaces and we showed how to obtain the Möbius strip, the torus, and the Klein bottle by making appropriate identifications of the edges of a rectangle. We propose to examine the construction of the Möbius strip in more detail and explain how to use the topology of the rectangle in order to make the Möbius strip into a topological space. The Möbius strip, when defined in this way, will be an example of an *identification space*.

The construction can be generalized, and the generalization will be the object of Section 4.2. The idea is to replace the rectangle by an arbitrary topological space X and to use the topology of X in order to make X, with certain of its points identified, into a topological space.

To construct a Möbius strip, one takes a rectangle and identifies a pair of opposite edges with a half twist. Our first job is to translate this process into precise mathematical language. For the rectangle take the subspace R of $\mathbb{E}^2$ consisting of those points (x,y) for which $0 \leqslant x \leqslant 3$ and $0 \leqslant y \leqslant 1$. To describe the identification of the vertical edges of R with a half twist, we partition R into disjoint nonempty subsets in such a way that two points lie in the same subset if and only if we wish them to be identified. If we now take these subsets as the *points* of our Möbius strip, then we have made the required identifications. The appropriate partition of R consists of:

(a) sets consisting of a pair of points of the form $(0,y)$, $(3, 1 - y)$, where $0 \leqslant y \leqslant 1$;

(b) sets consisting of a single point (x,y) where $0 < x < 3, 0 \leqslant y \leqslant 1$.

So far we have defined a set which we shall call M, its points being the subsets of the above partition of R. There is a natural function π from R onto M that sends each point of R to the subset of the partition in which it lies. The identification topology on M is defined to be the largest topology for which π is continuous. That is to say, a subset O is defined to be open in the identification topology on M if and only if $\pi^{-1}(O)$ is open in the rectangle R.

A glance at Fig. 4.1 shows the sort of open sets we obtain. We represent the points of M in the usual way as a subset of $\mathbb{E}^3$, and we label with the letter L

the image under π of the two vertical edges of R. If we use R_* to denote R minus its vertical edges, then the restriction of π to R_* is one–one and is a homeomorphism of R_* with $M - L$. Therefore we know all about the neighbourhoods

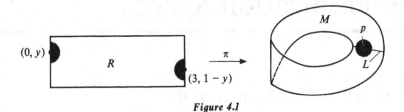

Figure 4.1

of points of $M - L$: they are simply the images under π of neighbourhoods of points of R_*. If p lies on the line L then $\pi^{-1}(p)$ consists of two distinct points, situated on the vertical edges of R, of the form $(0,y)$, $(3,1 - y)$. The union of two open half-discs† in R, centres $(0,y)$, $(3,1 - y)$ and of equal radius, maps via π to an open neighbourhood of p in the identification topology on M. Notice that if we take a single half-disc, its image in M is not a neighbourhood of p and is not open, so π is not an open mapping. The points of L are in no sense special in the Möbius strip; they have the same sort of neighbourhoods in the identification topology as all the other points of M. In fact, it is easy to check that the identification topology coincides with that induced from $\mathbb{E}^3$ on our set M.

For convenience we have illustrated M pictorially in $\mathbb{E}^3$. However, we emphasize that the definition of the Möbius strip as an 'identification space' given in this section is entirely abstract, and in no way relies on a particular representation of the strip as a set of points in euclidean space.

4.2 The identification topology

Let X be a topological space and let $\mathscr{P}$ be a family of disjoint nonempty subsets of X such that $\bigcup \mathscr{P} = X$. Such a family is usually called a partition of X. We form a new space Y, called an *identification space*, as follows. The points of Y are the members of $\mathscr{P}$ and, if $\pi : X \to Y$ sends each point of X to the subset of $\mathscr{P}$ containing it, the topology of Y is the largest for which π is continuous. Therefore a subset O of Y is open if and only if $\pi^{-1}(O)$ is open in X. This topology is called the *identification topology* on Y. We think of Y as the space obtained from X by identifying each of the subsets of $\mathscr{P}$ to a single point.

Our construction of the Möbius strip in Section 4.1 was a special case of this procedure. We shall give several other concrete examples below, but first we

---

† Quarter-discs if p is an endpoint of L.

prove one or two general results on identification spaces. We begin with a theorem that is useful when checking the continuity of a function which has an identification space as domain.

(4.1) Theorem. *Let* Y *be an identification space defined as above and let* Z *be an arbitrary topological space. A function* $f: Y \to Z$ *is continuous if and only if the composition* $f\pi: X \to Z$ *is continuous.*

Proof. Let U be an open subset of Z. Then $f^{-1}(U)$ is open in Y if and only if $\pi^{-1}(f^{-1}(U))$ is open in X, i.e., if and only if $(f\pi)^{-1}(U)$ is open in X.

Let $f: X \to Y$ be an onto map and suppose that the topology on Y is the largest for which f is continuous. Then we call f an *identification map*, the reason for our terminology being as follows. Any function $f: X \to Y$ gives rise to a partition of X whose members are the subsets $\{f^{-1}(y)\}$, where $y \in Y$. Let Y_* denote the identification space associated with this partition, and $\pi: X \to Y_*$ the usual map.

(4.2) Theorem. *If* f *is an identification map, then:*
(a) *the spaces* Y *and* Y_* *are homeomorphic;*
(b) *a function* $g: Y \to Z$ *is continuous if and only if the composition* $gf: X \to Z$ *is continuous.*

Proof. The proof of (b) is exactly that of theorem (4.1) because Y has the largest topology for which f is continuous. The points of Y_* are the sets $\{f^{-1}(y)\}$, where $y \in Y$. Define $h: Y_* \to Y$ by $h(\{f^{-1}(y)\}) = y$. Then h is a bijection and satisfies $h\pi = f, h^{-1}f = \pi$. By theorem (4.1), h is continuous, and h^{-1} is continuous by (b). Therefore h is a homeomorphism.

(4.3) Theorem. *Let* $f: X \to Y$ *be an onto map. If* f *maps open sets of* X *to open sets of* Y, *or closed sets to closed sets, then* f *is an identification map.*

Proof. Suppose f maps open sets to open sets. Let U be a subset of Y for which $f^{-1}(U)$ is open in X. Since f is onto, we have $f(f^{-1}(U)) = U$, and therefore U must be open in the given topology on Y. So this topology is the largest for which f is continuous, and f is an identification map. The proof for closed maps is similar.

(4.4) Corollary. *Let* $f: X \to Y$ *be an onto map. If* X *is compact and* Y *is Hausdorff, then* f *is an identification map.*

Proof. A closed subset of the compact space X is compact and its image under the continuous function f is therefore a compact subset of Y. But a compact subset of a Hausdorff space is closed. Therefore f takes closed sets to closed sets, and we can apply theorem (4.3).

We shall use theorem (4.2) and corollary (4.4) in order to compare different descriptions of the same topological space. We begin with two methods of constructing a torus.

The torus. Take X to be the unit square $[0,1] \times [0,1]$ in $\mathbb{E}^2$, with the subspace topology, and partition X into the following subsets:

(a) the set $\{(0,0),(1,0),(0,1),(1,1)\}$ of four corner points;
(b) sets consisting of pairs of points $(x,0)$, $(x,1)$, where $0 < x < 1$;
(c) sets consisting of pairs of points $(0,y)$, $(1,y)$, where $0 < y < 1$;
(d) sets consisting of a single point (x,y), where $0 < x < 1$ and $0 < y < 1$.

The resulting identification space is the torus. An equally common description is to say that the torus is the product $S^1 \times S^1$ of two circles. As usual, S^1 denotes the unit circle in the plane. Thinking of the points of S^1 as complex numbers, we can define a map $f:[0,1] \times [0,1] \to S^1 \times S^1$ by $f(x,y) = (e^{2\pi i x}, e^{2\pi i y})$. The partition of $[0,1] \times [0,1]$ which consists of the inverse images under f of points of $S^1 \times S^1$ is exactly that given earlier. By corollary (4.4), f is an identification map and therefore our two descriptions of the torus are homeomorphic.

The cone construction. We aim to define the cone on an arbitrary topological space X. Begin with $X \times I$ and let CX be the identification space associated with the partition which consists of:

(a) the subset $X \times \{1\}$;
(b) sets consisting of a single point (x,t), where $x \in X$ and $0 \leqslant t < 1$.

CX is called the *cone* on X. Intuitively we have pinched (identified) the top of $X \times I$ to a single point, this point becoming the apex of our cone.

If X happens to be a compact subspace of some euclidean space $\mathbb{E}^n$ there is an even more natural procedure. Include $\mathbb{E}^n$ in $\mathbb{E}^{n+1}$ as the set of points with final coordinate zero, and let v denote the point $(0,0,\ldots,0,1)$ of $\mathbb{E}^{n+1}$. Define the *geometric cone* on X to consist of those points of $\mathbb{E}^{n+1}$ which can be written in the form $tv + (1 - t)x$ where $x \in X$ and $0 \leqslant t \leqslant 1$. So the geometric cone is made up of all straight-line segments that join v to some point of X.

(4.5) Lemma. *The geometric cone on* X *is homeomorphic to* CX.

Proof. Define a function f from $X \times I$ to the geometric cone on X by $f(x,t) = tv + (1 - t)x$. Then f is continuous, onto, and $f(x,t) = f(x',t')$ if and only if either $x = x'$ and $t = t'$, or $t = t' = 1$. Therefore the partition of $X \times I$ induced by f is precisely that associated with the identification space CX. Since X is compact, $X \times I$ is also compact, and the geometric cone is of course Hausdorff since it lies in $\mathbb{E}^{n+1}$. Therefore f is an identification map by corollary (4.4) and the result follows from part (a) of theorem (4.2).

The identification space B^n/S^{n-1}. Let B^n denote the unit ball in n-dimensional euclidean space, and let S^{n-1} denote its boundary. Consider the partition of B^n which has as members:

(a) the set S^{n-1};
(b) the individual points of $B^n - S^{n-1}$.

The associated identification space is usually written B^n/S^{n-1}. In general, if we replace B^n by an arbitrary space X and S^{n-1} by a subspace A, then X/A means X with the subspace A identified to a point. Note that in this notation, CX becomes $X \times I/X \times \{1\}$.

We claim that B^n/S^{n-1} is homeomorphic to S^n. This is not very surprising. Take for example $n = 1$, then we are saying that identifying the endpoints of $[-1,1]$ gives a space homeomorphic to a circle. To give a formal proof we need only construct a map $f: B^n \rightarrow S^n$ which is onto, one–one on $B^n - S^{n-1}$, and which identifies all of S^{n-1} to a single point. Our map will be an identification map by corollary (4.4), and so theorem (4.2) provides the required homeomorphism. We can produce f as follows. We know that $\mathbb{E}^n$ is homeomorphic to $B^n - S^{n-1}$ and to $S^n - \{p\}$ for any point $p \in S^n$. Choose specific homeomorphisms $h_1: B^n - S^{n-1} \rightarrow \mathbb{E}^n$, $h_2: \mathbb{E}^n \rightarrow S^n - \{p\}$ and define

$$f(x) = \begin{cases} h_2 h_1(x) & \text{for} \quad x \in B^n - S^{n-1} \\ p & \text{for} \quad x \in S^{n-1} \end{cases}$$

The continuity of f is easy to check.

The glueing lemma. Let X, Y be subsets of a topological space and give each of X, Y, and $X \cup Y$ the induced topology. If $f: X \rightarrow Z$ and $g: Y \rightarrow Z$ are functions which agree on the intersection of X and Y, we can define

$$f \cup g: X \cup Y \rightarrow Z$$

by $f \cup g(x) = f(x)$ for $x \in X$, and $f \cup g(y) = g(y)$ for $y \in Y$. We say that $f \cup g$ is formed by 'glueing together' the functions f and g. The following result allows us, under certain conditions, to deduce the continuity of $f \cup g$ from the continuity of f and g.

(4.6) Glueing lemma. *If* X *and* Y *are closed in* X $\cup$ Y, *and if both* f *and* g *are continuous, then* f $\cup$ g *is continuous.*

Proof. Let C be a closed subset of Z. Then $f^{-1}(C)$ is closed in X (by the continuity of f), and therefore closed in $X \cup Y$ (since X is closed in $X \cup Y$). Similarly, $g^{-1}(C)$ is closed in $X \cup Y$. But $(f \cup g)^{-1}(C) = f^{-1}(C) \cup g^{-1}(C)$, and therefore $(f \cup g)^{-1}(C)$ is closed in $X \cup Y$. This proves $f \cup g$ is continuous.

The glueing lemma remains true if we ask that X and Y are both open in $X \cup Y$. We have stated the result for the closed case because it is this case that is most useful in practice. The lemma is of course false if we place no restrictions on X and Y.

69

As we shall see, the glueing lemma can be explained in terms of identification maps and interpreted as a special case of theorem (4.3). In order to do this, we introduce the *disjoint* union $X + Y$ of the spaces X, Y, and the function $j : X + Y \to X \cup Y$ which when restricted to either X or Y is just the inclusion in $X \cup Y$. This function is important for our purposes because:

(a) it is continuous;
(b) the composition $(f \cup g)j : X + Y \to Z$ is continuous if and only if both f and
    g are continuous.

By combining (b) and part (b) of theorem (4.2), we have the following result:

(4.7) Theorem. *If j is an identification map, and if both* $f : X \to Z$ *and* $g : Y \to Z$ *are continuous, then* $f \cup g : X \cup Y \to Z$ *is continuous.*

The glueing lemma is a special case of this result, since if both X and Y are closed in $X \cup Y$, then j sends closed sets to closed sets and is an identification map by theorem (4.3).

If j is an identification map, then we can think of $X \cup Y$ as an identification space formed from the disjoint union $X + Y$ by identifying certain points of X with points of Y. In this case, we often say that $X \cup Y$ has the identification topology. The open (closed) sets of $X \cup Y$ are those sets A for which $A \cap X$ and $A \cap Y$ are open (closed) in X and Y respectively.

Theorem (4.7) generalizes to the case of an arbitrary union. Let $X_\alpha, \alpha \in A$, be a family of subsets of a topological space and give each X_α, and the union $\bigcup X_\alpha$, the induced topology. Let Z be a space and suppose we are given maps $f_\alpha : X_\alpha \to Z$, one for each α in A, such that if $\alpha, \beta \in A$,

$$f_\alpha \,|\, X_\alpha \cap X_\beta = f_\beta \,|\, X_\alpha \cap X_\beta$$

Define a function $F : \bigcup X_\alpha \to Z$ by glueing together the f_α, i.e., $F(x) = f_\alpha(x)$ if $x \in X_\alpha$. Let $\oplus X_\alpha$ denote the disjoint union of the spaces X_α, and let $j : \oplus X_\alpha \to \bigcup X_\alpha$ be the function which when restricted to each X_α is the inclusion in $\bigcup X_\alpha$.

(4.8) Theorem. *If j is an identification map, and if each* f_α *is continuous, then F is continuous.*

Proof. Observe that $Fj : \oplus X_\alpha \to Z$ is continuous if and only if each f_α is continuous, and apply part (b) of theorem (4.2).

As before, we say that $\bigcup X_\alpha$ has the identification topology when j is an identification map. If the X_α are finite in number, and if each X_α is closed in $\bigcup X_\alpha$, then $\bigcup X_\alpha$ automatically has the identification topology. If the X_α are infinite in number, one must be careful. Figure 4.2 represents an infinite collection of closed intervals in the plane. The subspace topology on their union quite clearly gives a space homeomorphic to the circle, whereas the identification topology gives a space homeomorphic to the nonnegative part of the real line (send the interval labelled n to $[n - 1, n]$).

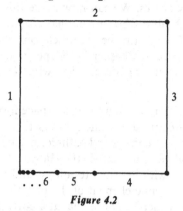

Figure 4.2

Projective spaces. We give three descriptions of real n-dimensional projective space P^n. As usual, theorem (4.2) and corollary (4.4) can be used to show that all three lead to the same space.

(a) Take the unit sphere S^n in $\mathbb{E}^{n+1}$ and partition it into subsets which contain exactly two points, the points being antipodal (at opposite ends of a diameter). P^n is the resulting identification space. We could abbreviate our description by saying that P^n is formed from S^n by identifying antipodal points.

(b) Begin with $\mathbb{E}^{n+1} - \{0\}$ and identify two points if and only if they lie on the same straight line through the origin. (Note that antipodal points of S^n have this property.)

(c) Begin with the unit ball B^n and identify antipodal points of its boundary sphere.

Attaching maps. As a final example of an identification space we formalize the notion of attaching one space to another by means of a continuous function.

Let X, Y be spaces, let A be a subspace of Y, and let $f : A \rightarrow X$ be a continuous function. Our aim is to attach Y to X using f and to form a new space which we shall denote by $X \cup_f Y$. We begin with the disjoint union $X + Y$ and define a partition so that two points lie in the same subset if and only if they are identified under f. Precisely, the subsets of the partition are:

(a) pairs of points $\{a, f(a)\}$ where $a \in A$;
(b) individual points of $Y - A$;
(c) individual points of $X - \text{image}(f)$.

The identification space associated with this partition is $X \cup_f Y$. The map f is called the *attaching map*.

In many applications, Y will be a ball and A its boundary sphere. Consider

71

the description of the projective plane (real projective space of dimension 2) given in Chapter 1. The idea was to attach a disc to a Möbius strip by glueing together their boundary circles. We can now make this precise. Let M denote the Möbius strip and D the disc. Choose a homeomorphism h from the boundary circle of D to that of M and form the identification space $M \cup_h D$. The result is P^2 and is (as we shall see in Chapter 7) independent of the choice of h. We leave the reader to reconcile this description with those listed in 'Projective spaces' above.

One final comment: if Y is an identification space formed from X, then Y is the image of X under a continuous function and therefore inherits properties such as compactness, connectedness, and path-connectedness from X. However, X may be Hausdorff and yet Y not satisfy the Hausdorff axiom. As an example, take X to be the real line with its usual topology, and partition X so that real numbers r and s lie in the same element of the partition if and only if $r - s$ is rational. We invite the reader to check that the corresponding identification space is an *indiscrete* space.

Problems

1. Check that the three descriptions (a), (b), (c) of P^n listed in 'Projective spaces' above do all lead to the same space.

2. Which space do we obtain if we take a Möbius strip and identify its boundary circle to a point?

3. Let $f:X \to Y$ be an identification map, let A be a subspace of X, and give $f(A)$ the induced topology from Y. Show that the restriction $f|A:A \to f(A)$ need not be an identification map.

4. With the terminology of Problem 3, show that if A is open in X and if f takes open sets to open sets, or if A is closed in X and f takes closed sets to closed sets, then $f|A:A \to f(A)$ is an identification map.

5. Let X denote the union of the circles $[x - (1/n)]^2 + y^2 = (1/n)^2, n = 1,2,3,\ldots$, with the subspace topology from the plane, and let Y denote the identification space obtained from the real line by identifying all the integers to a single point. Show that X and Y are not homeomorphic. (X is called the *Hawaiian earring*.)

6. Give an example of an identification map which is neither open nor closed.

7. Describe each of the following spaces: (a) the cylinder with each of its boundary circles identified to a point; (b) the torus with the subset consisting of a meridianal and a longitudinal circle identified to a point; (c) S^2 with the equator identified to a point; (d) $\mathbb{E}^2$ with each of the circles centre the origin and of integer radius identified to a point.

8. Let X be a compact Hausdorff space. Show that the cone on X is homeomorphic to the one-point compactification of $X \times [0,1)$. If A is closed in X, show that X/A is homeomorphic to the one-point compactification of $X - A$.

9. Let $f : X \to X'$ be a continuous function and suppose we have partitions $\mathscr{P}, \mathscr{P}'$ of X and X' respectively, such that if two points of X lie in the same member of $\mathscr{P}$, their images under f lie in the same member of $\mathscr{P}'$. If Y, Y' are the identification spaces given by these partitions, show that f induces a map $\hat{f} : Y \to Y'$, and that if f is an identification map then so is $\hat{f}$.

10. Let S^2 be the unit sphere in $\mathbb{E}^3$ and define $f : S^2 \to \mathbb{E}^4$ by $f(x,y,z) = (x^2 - y^2,$ $xy, xz, yz)$. Show that f induces an embedding of the projective plane in $\mathbb{E}^4$ (embeddings were defined in Problem 14 of Chapter 3).

11. Show that the function $f : [0,2\pi] \times [0,\pi] \to \mathbb{E}^5$ defined by $f(x,y) = (\cos x,$ $\cos 2y, \sin 2y, \sin x \cos y, \sin x \sin y)$ induces an embedding of the Klein bottle in $\mathbb{E}^5$.

12. With the notation of Problem 11, show that if $(2 + \cos x)\cos 2y = (2 + \cos x')\cos 2y'$ and $(2 + \cos x)\sin 2y = (2 + \cos x')\sin 2y'$, then $\cos x = \cos x'$, $\cos 2y = \cos 2y'$, and $\sin 2y = \sin 2y'$. Deduce that the function $g : [0,2\pi] \times [0,\pi] \to \mathbb{E}^4$ given by $g(x,y) = ((2 + \cos x)\cos 2y, (2 + \cos x)\sin 2y,$ $\sin x \cos y, \sin x \sin y)$ induces an embedding of the Klein bottle in $\mathbb{E}^4$.

4.3 Topological groups

We leave the notion of an identification space briefly in order to consider spaces which have, in addition to their topology, the structure of a group. A good example is the circle, thought of as the set of complex numbers of unit modulus. Its topology is that induced from the plane and the group structure is simply multiplication of complex numbers. Note that the two functions

$$S^1 \times S^1 \to S^1$$
$$(e^{i\theta}, e^{i\phi}) \mapsto e^{i(\theta + \phi)} \qquad \textit{(group multiplication)}$$

$$S^1 \to S^1$$
$$e^{i\theta} \mapsto e^{-i\theta} \qquad \textit{(inversion in the group)}$$

are continuous, so the topology and the algebraic structure fit together nicely.

(4.9) Definition. *A topological group* G *is both a Hausdorff topological space and a group, the two structures being compatible in the sense that the group multiplication* m : G × G → G, *and the function* i : G → G *which sends each group element to its inverse, are continuous.*

Most of this section will be taken up by examples, including examples of matrix groups. In Section 4.4 we return to identification spaces. We shall define there the action of a topological group on a space, show how an action leads to an identification space, and consider a variety of identification spaces which arise in this way.

Examples of topological groups

1. The real line, the group structure being addition of real numbers.

2. The circle, as described above.

3. Any abstract group with the discrete topology.

4. The torus considered as the product of two circles. We take the product topology and the product group structure. (The product of two topological groups is a topological group; see Problem 13.)

5. The three-sphere considered as the unit sphere in the space of quaternions $\mathbb{H}$. ($\mathbb{H}$ is topologically $\mathbb{E}^4$ and has the algebraic structure of the quaternions.)

6. Euclidean n-space. We choose the notation $\mathbb{R}^n$ to emphasize that we have a topological group (usual addition as group structure) and not simply the topological space $\mathbb{E}^n$.

7. The group of invertible $n \times n$ matrices with real entries. The group structure is matrix multiplication. For the topology we identify each $n \times n$ matrix $A = (a_{ij})$ with the corresponding point

$$(a_{11}, a_{12}, \ldots, a_{1n}, a_{21}, \ldots, a_{2n}, a_{31}, \ldots, a_{nn})$$

of $\mathbb{E}^{n^2}$ and take the subspace topology. This topological group is called the *general linear group*, and we denote it by $GL(n)$.[†] A detailed verification that $GL(n)$ is a topological group will be given in theorem (4.12).

8. The *orthogonal group* $O(n)$ consisting of $n \times n$ orthogonal matrices with real entries. $O(n)$ has both its topology and its group structure induced from $GL(n)$. It is a subgroup (as a topological group) of $GL(n)$. The subgroup of $O(n)$ consisting of those matrices which have determinant $+1$ is called the *special orthogonal group* and written $SO(n)$.

The terms 'isomorphism' and 'subgroup' for topological groups require a few words of explanation. In each case we need to take into consideration both the topological and the algebraic structures. So an isomorphism between two topological groups is a homeomorphism which is also a group isomorphism. In the same spirit, a subset of a topological group is called a subgroup if it is algebraically a subgroup and in addition has the subspace topology. Therefore the integers $\mathbb{Z}$ with the discrete topology form a subgroup of the real line $\mathbb{R}$. If we form the factor group $\mathbb{R}/\mathbb{Z}$ and give it the identification topology (the corresponding partition of $\mathbb{R}$ is that given by the cosets of $\mathbb{Z}$) then we have a topological group isomorphic to the circle. For the map $f : \mathbb{R} \to S^1$ defined by $f(x) = e^{2\pi i x}$ takes open sets to open sets and is an identification map, by theorem (4.3). Two points of $\mathbb{R}$ are identified by f if and only if they differ by an

---

† Or $GL(n, \mathbb{R})$ to emphasize that the matrices have real entries. $GL(n, \mathbb{C})$ then denotes the corresponding group of invertible matrices with complex entries.

integer, and therefore f induces a homeomorphism of $\mathbb{R}/\mathbb{Z}$† with S^1, by theorem (4.2). It is elementary to check that this homeomorphism is a group isomorphism. For a second example involving the ideas of subgroup and isomorphism, we turn to our matrix groups. Associating each $(n-1) \times (n-1)$ orthogonal matrix A with the $n \times n$ orthogonal matrix

$$\begin{pmatrix} 1 & 0 \\ 0 & A \end{pmatrix}$$

shows that $O(n-1)$ is isomorphic to a subgroup of $O(n)$

Let G be a topological group and x an element of G. The function $L_x : G \to G$ defined by $L_x(g) = xg$ is called *left translation* by the element x. It is clearly one–one and onto, and it is continuous because it is the composition

$$G \to G \times G \overset{m}{\to} G$$

$$g \mapsto (x,g) \mapsto xg.$$

The inverse of L_x is $L_{x^{-1}}$ and therefore L_x is a homeomorphism. Similarly the right translation $R_x : G \to G$ given by $R_x(g) = gx$ is also a homeomorphism.

These translations show that a topological group has a certain 'homogeneity' as a topological space. For if x and y are any two points of a topological group G there is a homeomorphism of G that maps x to y, namely the translation $L_{yx^{-1}}$. Therefore G exhibits the same topological structure locally near each point.

(4.10) Theorem. *Let* G *be a topological group and let* K *denote the connected component of* G *which contains the identity element. Then* K *is a closed normal subgroup of* G.
Remark. If $G = O(n)$ then $K = SO(n)$. We shall prove this later.

Proof. Components are always closed. For any $x \in K$ the set $Kx^{-1} = R_{x^{-1}}(K)$ is connected (since $R_{x^{-1}}$ is a homeomorphism) and contains $e = xx^{-1}$. Since K is the maximal connected subset of G containing e, we must have $Kx^{-1} \subseteq K$. Therefore $KK^{-1} = K$, and K is a subgroup of G. Normality follows in a similar manner. For any $g \in G$ the set $gKg^{-1} = R_{g^{-1}}L_g(K)$ is connected and contains e. Therefore $gKg^{-1} \subseteq K$.

(4.11) Theorem. *In a connected topological group any neighbourhood of the identity element is a set of generators for the whole group.*

---

† We have an unfortunate clash of notation. $\mathbb{R}/\mathbb{Z}$ is used for the identification space whose points are the cosets of $\mathbb{Z}$ in $\mathbb{R}$, and for $\mathbb{R}$ with the single subspace $\mathbb{Z}$ pinched to a point. The first of these is the circle, the second is an infinite bouquet of circles (i.e., an infinite collection of circles all joined together at one point). It should always be clear from the context which of the possibilities we are considering.

Proof. Let G be a connected topological group and let V be a neighbourhood of e in G. Let $H = \langle V \rangle$ be the subgroup of G generated by the elements of V. If $h \in H$ then the whole neighbourhood $hV = L_h(V)$ of h lies in H, so H is open. We claim that the complement of H is also open. For if $g \in G - H$, consider the set gV. If $gV \cap H$ is nonempty, say $x \in gV \cap H$, then $x = gv$ for some $v \in V$. This gives $g = xv^{-1}$, which implies the contradiction $g \in H$ since both x and v^{-1} lie in H. Therefore the neighbourhood $L_g(V) = gV$ of g lies in $G - H$, and we see that $G - H$ is an open set. Now G is connected and so cannot be partitioned into two disjoint nonempty open sets. Since H is nonempty we must have $G - H = \varnothing$, i.e., $G = H$.

(4.12) Theorem. *The matrix group* GL(n) *is a topological group.*

Proof. Let M denote the set of all $n \times n$ matrices which have real entries, and let $A = (a_{ij})$ represent a typical element of M. We can identify M with euclidean space of dimension n^2 by associating $A = (a_{ij})$ with the point $(a_{11}, a_{12}, \ldots, a_{1n}, a_{21}, \ldots, a_{2n}, a_{31}, \ldots, a_{nn})$. The identification gives us a topology on M and we claim that, with respect to this topology, matrix multiplication $m : \mathsf{M} \times \mathsf{M} \to \mathsf{M}$ is continuous. To see this, we need only examine the well-known formula for the entries of a product matrix: if $A = (a_{ij})$ and $B = (b_{ij})$ then the ijth entry in the product $m(A, B)$ is $\sum_{k=1}^{n} a_{ik} b_{kj}$. Now M has the topology of the product space $\mathbb{E}^1 \times \mathbb{E}^1 \times \ldots \times \mathbb{E}^1$ (n^2 copies), and for each i, j satisfying $1 \leqslant i, j \leqslant n$ we have a projection $\pi_{ij} : \mathsf{M} \to \mathbb{E}^1$ which sends a given matrix A to its ijth entry. By theorem (3.13), m is continuous if and only if all of the composite functions

$$\mathsf{M} \times \mathsf{M} \xrightarrow{\ m\ } \mathsf{M} \xrightarrow{\ \pi_{ij}\ } \mathbb{E}^1$$

are continuous. But $\pi_{ij} m(A, B) = \sum_{k=1}^{n} a_{ik} b_{kj}$, a polynomial in the entries of A and B. Therefore $\pi_{ij} m$ is continuous.

The elements of GL(n) are the *invertible* matrices in M. If we give GL(n) the subspace topology from M then, by the above, matrix multiplication GL(n) $\times$ GL(n) $\to$ GL(n) is continuous. It remains to prove that the inverse function $i : \text{GL}(n) \to \text{GL}(n)$ is also continuous. We use the same technique: $i : \text{GL}(n) \to$ GL(n) $\subseteq \mathbb{E}^1 \times \ldots \times \mathbb{E}^1$ is continuous if and only if all of the composite functions

$$\text{GL}(n) \xrightarrow{\ i\ } \text{GL}(n) \xrightarrow{\ \pi_{jk}\ } \mathbb{E}^1 \qquad 1 \leqslant j, k \leqslant n$$

are continuous. Now the composition of π_{jk} with i sends a matrix A to the jkth element of A^{-1}, i.e., to $(1/\det A)$ (kjth cofactor of A). It should be clear that the determinant of A and the cofactors of A are polynomials in the entries of A. Since $\det A$ does not vanish on GL(n), our composition $\pi_{jk} i$ is continuous. This completes the proof that GL(n) is a topological group.

We note in passing that GL(n) is the inverse image of the nonzero real numbers under the determinant function $\det: M \to \mathbb{R}$. So GL(n) *is not compact* (it is an open subset of M), and is *not connected* (the matrices with positive and negative determinants partition GL(n) into two disjoint nonempty open sets). How many components has GL(n)?

(4.13) Theorem. O(n) *and* SO(n) *are compact*.

Proof. O(n) consists of those matrices in GL(n) which have their transpose as inverse. It is algebraically a subgroup of GL(n) and we give it the subspace topology. In order to show O(n) compact we show that it corresponds to a closed bounded subset of $\mathbb{E}^{n^2}$ under our identification of M with $\mathbb{E}^{n^2}$.

Let $A \in$ O(n). Since $AA^t = I$ we have $\sum_{j=1}^{n} a_{ij}a_{kj} = \delta_{ik}$ for $1 \leqslant i,k \leqslant n$. For each choice of i,k we define a map $f_{ik}: M \to \mathbb{E}^1$ by $f_{ik}(A) = \sum_{j=1}^{n} a_{ij}a_{kj}$. Then O($n$) is the intersection of all sets of the form

$$f_{ik}^{-1}(0) \qquad 1 \leqslant i,k \leqslant n, \qquad i \neq k$$

$$f_{ii}^{-1}(1) \qquad 1 \leqslant i \leqslant n$$

Therefore O(n) is closed in M since it is the intersection of a finite number of closed sets.

For the boundedness of O(n) we have only to look at the conditions $\sum_{j=1}^{n} a_{ij}a_{ij} = 1$. These imply that the entries of any orthogonal matrix A satisfy $|a_{ij}| \leqslant 1$. This completes the proof that O(n) is compact.

Finally, SO(n) is compact because it is closed in O(n).

We note that SO(2) $\cong S^1$, and SO(3) $\cong P^3$, where $\cong$ means isomorphism of topological groups. Sending the rotation matrix

$$\begin{pmatrix} \cos\theta & -\sin\theta \\ \sin\theta & \cos\theta \end{pmatrix}$$

to the point $e^{i\theta}$ of S^1 gives the first of these. For the second, we think of S^3 as the quaternions of norm 1, and note that conjugation in $\mathbb{H}$ by a nonzero quaternion always induces a rotation of the three-dimensional subspace of pure quaternions. This defines a function $\mathbb{H} - \{0\} \to$ SO(3) which is in fact (check these statements!) a homomorphism, onto, and continuous. Its kernel is $\mathbb{R} - \{0\}$. Restricting this function to S^3 gives a continuous epimorphism from S^3 to SO(3) with kernel $\{1, -1\}$. Now the set of cosets $S^3/\{1, -1\}$, with the identification topology, is of course P^3, and therefore we have a continuous group isomorphism $P^3 \to$ SO(3). Since P^3 is compact and SO(3) is Hausdorff, this map is a homeomorphism.

77

Problems

13. Show that the product of two topological groups is a topological group.

14. Let G be a topological group. If H is a subgroup of G, show that its closure $\bar{H}$ is also a subgroup, and that if H is normal then so is $\bar{H}$.

15. Let G be a compact Hausdorff space which has the structure of a group. Show that G is a topological group if the multiplication function $m : G \times G \to G$ is continuous.

16. Prove that $O(n)$ is homeomorphic to $SO(n) \times Z_2$. Are these two isomorphic as topological groups?

17. Let A,B be compact subsets of a topological group. Show that the product set $AB = \{ab \mid a \in A, b \in B\}$ is compact.

18. If U is a neighbourhood of e in a topological group, show there is a neighbourhood V of e for which $VV^{-1} \subseteq U$.

19. Let H be a discrete subgroup of a topological group G (i.e., H is a subgroup, and is a discrete space when given the subspace topology). Find a neighbourhood N of e in G such that the translates $hN = L_h(N)$, $h \in H$, are all disjoint.

20. If C is a compact subset of a topological group G, and if H is a discrete subgroup of G, show that $H \cap C$ is finite.

21. Prove that every nontrivial discrete subgroup of $\mathbb{R}$ is infinite cyclic.

22. Prove that every nontrivial discrete subgroup of the circle is finite and cyclic.

23. Let $A,B \in O(2)$ and suppose $\det A = +1$, $\det B = -1$. Show that $B^2 = I$ and $BAB^{-1} = A^{-1}$. Deduce that every discrete subgroup of $O(2)$ is either cyclic or dihedral.

24. If T is an automorphism of the topological group $\mathbb{R}$ (i.e., T is a homeomorphism which is also a group isomorphism) show that $T(r) = rT(1)$ for any rational number r. Deduce that $T(x) = xT(1)$ for any real number x, and hence that the automorphism group of $\mathbb{R}$ is isomorphic to $\mathbb{R} \times \mathbb{Z}_2$.

25. Show that the automorphism group of the circle is isomorphic to $\mathbb{Z}_2$.

4.4 Orbit spaces

The infinite cyclic group $\mathbb{Z}$ can be thought of as a group of homeomorphisms of the real line in a very natural way. Each integer $n \in \mathbb{Z}$ determines a translation $x \mapsto x + n$ of the line.

If we consider the matrix group $O(n)$, then each matrix gives rise to a linear transformation of euclidean n-space. Since the elements of $O(n)$ are invertible, and since orthogonal transformations preserve the euclidean metric (and therefore send unit vectors to unit vectors), each orthogonal matrix gives us a homeomorphism from the unit sphere S^{n-1} to itself. This operation of the orthogonal group on the sphere is compatible with the topologies of $O(n)$ and S^{n-1} in the sense that the function

$$O(n) \times S^{n-1} \to S^{n-1}$$

$$(A,\mathbf{x}) \mapsto A\mathbf{x}$$

is continuous. We say that $O(n)$ 'acts' on the space S^{n-1} as a group of homeomorphisms.

If we give $\mathbb{Z}$ its natural topology (the discrete topology induced from $\mathbb{R}$), then both of these examples fit into a general setting.

(4.14) Definition. *A topological group* G *is said to act as a group of homeomorphisms on a space* X *if each group element induces a homeomorphism of the space in such a way that:*

(a) hg(x) = h(g(x))† *for all* g,h ∈ G, *for all* x ∈ X;
(b) e(x) = x *for all* x ∈ X, *where* e *is the identity element of* G;
(c) *the function* G × X → X *defined by* (g,x) ↦ g(x) *is continuous.*

If x is a point of the space X, then for each $g \in G$ the corresponding homeomorphism either fixes x or maps it to some new point $g(x)$. The subset of X consisting of all such images $g(x)$, as g varies through G, is called the *orbit* of x and written $O(x)$. If two orbits intersect then they must coincide: the relation defined by $x \sim y$ if and only if $x = g(y)$ for some $g \in G$ is an equivalence relation on X whose equivalence classes are precisely the orbits of the given action. So the orbits define a partition of X. The corresponding identification space is called the *orbit space* and is written X/G. In constructing X/G we 'divide' by G in the sense that we identify two points of X if and only if they differ by one of the homeomorphisms $x \mapsto g(x)$.

In our first example, the orbit of a real number x consists of all points $x + n$ where $n \in \mathbb{Z}$. Therefore in forming $\mathbb{R}/\mathbb{Z}$ we identify two points of $\mathbb{R}$ if and only if they differ by an integer and, as explained in the preceding section, we obtain the circle as orbit space.

The orthogonal action on S^{n-1} is an example of a *transitive action*, that is, an action for which the orbit of any point is the whole space (in this case all of S^{n-1}). The proof is quite easy. Let $e_1, e_2, \ldots, e_n$ be the standard orthonormal basis for $\mathbb{E}^n$ and, given $x \in S^{n-1}$, construct a second orthonormal basis with x as first member. If A is the matrix of this new basis with respect to $e_1, e_2, \ldots, e_n$, then A is orthogonal and $A(e_1) = x$. Therefore we have shown that the orbit of e_1 is all of S^{n-1}. Whenever we have a transitive action, i.e., only one distinct orbit, then of course the orbit space is a single point.

---

† We use the same letter for a group element and the homeomorphism induced by it.

More examples

1. Taking the product of our first example with itself in the natural way gives an action of $\mathbb{Z} \times \mathbb{Z}$ on the plane. An ordered pair of integers $(m,n) \in \mathbb{Z} \times \mathbb{Z}$ sends the point $(x,y) \in \mathbb{E}^2$ to $(x + m, y + n)$. The orbit space is the product of two circles, in other words the torus. It may help to think of this action geometrically. Divide the plane into squares of unit side by drawing in all horizontal and vertical lines through the points with integer coordinates. The homeomorphisms of our group action preserve this pattern of squares, and any single square contains points from each orbit and therefore maps *onto* the torus under the identification map

$$\mathbb{E}^2 \xrightarrow{\;\pi\;} \mathbb{E}^2/\mathbb{Z} \times \mathbb{Z} = T$$

Each square has its sides identified by π in the usual way in order to form T.

2. We describe an action of $\mathbb{Z}_2$ on the n-sphere which has the projective space P^n as orbit space. $\mathbb{Z}_2$ has only two elements.† We know from the definition of a group action that the identity element must give rise to the identity homeomorphism, and we ask that the generator (i.e., the non-identity element) give the *antipodal map* which takes each point of S^n to its antipode. (Note that if we do this homeomorphism twice then we arrive at the identity homeomorphism of S^n.) The orbits of the action are pairs of antipodal points and the orbit space corresponds to one of our descriptions of P^n given in Section 4.2.

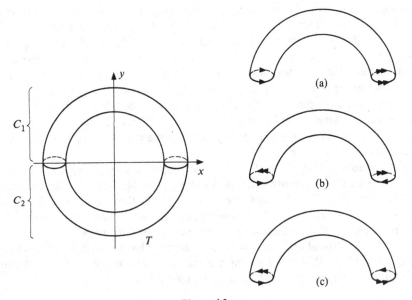

Figure 4.3

---

† We take the discrete topology when the group is finite.

3. A given group may act in many different ways on the same space. Here are three different actions of $\mathbb{Z}_2$ on the torus. Take the torus T in $\mathbb{E}^3$ formed by rotating the circle $(x - 3)^2 + z^2 = 1$ about the z axis. Let g denote the generator of $\mathbb{Z}_2$ and define:

(a) $g(x,y,z) = (x, -y, -z)$, rotation of T through π about the x axis;
(b) $g(x,y,z) = (-x, -y, z)$, rotation of T through π about the z axis;
(c) $g(x,y,z) = (-x, -y, -z)$, reflexion of T in the origin.

Each of these homeomorphisms determines an action of $\mathbb{Z}_2$ on T. The orbit spaces are the sphere, torus, and Klein bottle respectively, and Fig. 4.3 shows why. In each case g interchanges the cylinders C_1 and C_2. So in order to form the orbit space we can ignore C_2 and simply make the appropriate identifications on the boundary circles of C_1.

4. If G is a topological group and H a subgroup of G, then H acts on G by left translation. The homeomorphism induced by an element h of H is L_h, i.e., $h(g) = L_h(g) = hg$, and the associated function $H \times G \to G$ is continuous, since multiplication in G is continuous. Two elements g, g' lie in the same orbit if and only if $g' \in Hg$. Therefore the orbits are the right cosets of H in G.

We also have a 'right action' of H on G given by the map

$$H \times G \to G$$

$$(h,g) \mapsto R_{h^{-1}}(g)$$

where the inversion of h is needed to make property (a) of definition (4.14) valid. The orbits are now the left cosets of H in G.

We denote both orbit spaces by G/H; they are of course homeomorphic.

5. We return to the action of $O(n)$ on S^{n-1}. Note that if $A \in O(n)$, and if $A(e_1) = e_1$, then A has the form

$$\begin{pmatrix} 1 & 0 \\ 0 & B \end{pmatrix}$$

where B is orthogonal. Conversely any matrix of this form fixes e_1. Therefore the subgroup† of $O(n)$ consisting of those elements which leave e_1 fixed is isomorphic to $O(n - 1)$.

We can define a function $f : O(n) \to S^{n-1}$ by $f(A) = A(e_1)$. This function is continuous because it is the composition

$$O(n) \to O(n) \times S^{n-1} \to S^{n-1}$$

$$A \mapsto (A, e_1) \mapsto A(e_1)$$

and it is onto because the given action is transitive. Now $O(n)$ is compact and S^{n-1} Hausdorff, which makes f an identification map by corollary (4.4). If

---

† Often called the isotropy subgroup or stabilizer of e_1. Points in the same orbit always have conjugate stabilizers.

$\mathbf{x} \in S^{n-1}$ one easily checks that $f^{-1}(\mathbf{x})$ is precisely the left coset $A\,O(n-1)$, where $A \in O(n)$ satisfies $A(\mathbf{e}_1) = \mathbf{x}$. Therefore the partition of $O(n)$ induced by f coincides with the left-coset decomposition of $O(n)$ corresponding to the subgroup $O(n-1)$. Applying theorem (4.2) shows that $O(n)/O(n-1)$ *is homeomorphic to* S^{n-1}. A similar argument gives $SO(n)/SO(n-1) \cong S^{n-1}$.

We deduce, by induction, that $SO(n)$ is connected. The induction starts since $SO(1)$ is a single point. For the inductive step we use $SO(n+1)/SO(n) \cong S^n$ (remembering that the n-sphere is connected for $n \geqslant 1$) and the following theorem.

(4.15) Theorem. *Let* G *act on* X *and suppose that both* G *and* X/G *are connected, then* X *is connected.*

Proof. Suppose X is the union of two disjoint nonempty open subsets U and V. Since the identification map $\pi : X \to X/G$ always takes open sets to open sets (Problem 29), and since X/G is connected, $\pi(U)$ and $\pi(V)$ cannot be disjoint. Now if $\pi(x) \in \pi(U) \cap \pi(V)$, then both $U \cap O(x)$ and $V \cap O(x)$ are nonempty. These two sets decompose the orbit $O(x)$ as a disjoint union of two nonempty open sets. But $O(x)$ is the image of G under the continuous function $f : G \to X$ defined by $f(g) = g(x)$. $O(x)$ is therefore connected, and we have established the required contradiction.

6. Let p and q be relatively prime integers (not necessarily primes). Consider the 3-sphere as the unit sphere in complex space of dimension 2, that is

$$S^3 = \{(z_0, z_1) \in \mathbb{C}^2 \mid z_0 \bar{z}_0 + z_1 \bar{z}_1 = 1\}.$$

Let g denote the generator of the cyclic group $\mathbb{Z}_p$ and define an action of $\mathbb{Z}_p$ on S^3 by

$$g(z_0, z_1) = (e^{2\pi i/p} z_0, \, e^{2\pi q i/p} z_1).$$

Of course, having specified the effect of g, the homeomorphisms induced by $g^2, g^3, \ldots$ are completely determined by property (a) of definition (4.14) for a group action. If we repeat g a total of p times, we arrive at the identity homeomorphism. The quotient space $S^3/\mathbb{Z}_p$ is called a *Lens space* and written $L(p,q)$. We shall see later that $L(p,q)$ is locally euclidean of dimension 3, and has fundamental group isomorphic to $\mathbb{Z}_p$. (For an alternative description of $L(p,q)$, see Problem 33.)

7. So far, most of our orbits have been rather simple, namely discrete sets of points or the whole space. To show things can be much more complicated, we describe an action of the real line on the torus where each orbit is a dense proper subset of the torus. Identify the torus with $S^1 \times S^1$ and define the homeomorphism induced by the real number r to be

$$(e^{2\pi i x}, e^{2\pi i y}) \mapsto (e^{2\pi i(x+r)}, e^{2\pi i(y + r\sqrt{2})})$$

If $\pi : \mathbb{E}^2 \to S^1 \times S^1$ denotes the identification map $(x,y) \mapsto (e^{2\pi i x}, e^{2\pi i y})$, the orbits of this action are simply the images under π of the straight lines in the plane with gradient $\sqrt{2}$. The important fact for our purposes is that $\sqrt{2}$ is irrational. Notice that π is one–one when restricted to a line of gradient $\sqrt{2}$, since π can only identify $(x + r, y + r\sqrt{2})$, $(x + s, y + s\sqrt{2})$ if both $r - s$ and $r\sqrt{2} - s\sqrt{2}$ are integers, which is clearly impossible.

We shall examine the orbit of the point $\pi(0,0) \in T$. It is simply the image under π of the straight line through the origin (or through any other point with integer coordinates) in $\mathbb{E}^2$ with gradient $\sqrt{2}$. Call this line L. We can represent our orbit on the unit square in the plane (Fig. 4.4), remembering that the torus is formed from this square by identifying its edges in the usual way. If we travel away from the origin along L in the first quadrant, we stay inside the unit square as far as the point $(1/\sqrt{2}, 1)$. This point represents the same point on the torus as $(1/\sqrt{2}, 0)$ and we continue along our orbit with gradient $\sqrt{2}$ from $(1/\sqrt{2}, 0)$ to $(1, \sqrt{2} - 1)$. Now we jump (in the square, though not in the torus!) to $(0, \sqrt{2} - 1)$ and continue with gradient $\sqrt{2}$, etc.

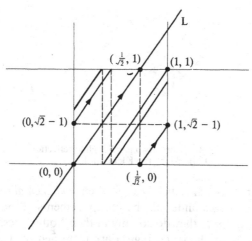

Figure 4.4

Our orbit winds round and round the torus. It almost fills out the whole torus, but not quite. We leave the reader to check for himself that the orbit is a proper dense subset of T.

This action of $\mathbb{R}$ on T is called an 'irrational flow' on the torus, the orbits being called 'flow lines'.

8. We end this section by mentioning a rather interesting class of groups of isometries of the plane. The groups we shall consider have the property that they preserve some repeating pattern of convex polygons which fills out the whole plane (i.e., the elements of the group are all symmetries of the pattern). We illustrate three examples in Fig 4.5 by giving in each case a set of generators for the group. We shall denote the magnitude and direction of a *translation* by an arrow $\longrightarrow$. A half-arrow $\longrightarrow$ will represent a *glide reflection*, that is a reflection in the line of the arrow followed by a translation of magnitude and direction indicated by the arrow. Rotation through 180° about the midpoint

of a line segment will be denoted by 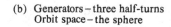 and called a half-turn.

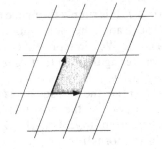

(a) Generators – two translations
    Orbit space – the torus

(b) Generators – three half-turns
    Orbit space – the sphere

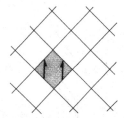

Figure 4.5

(c) Generators – two parallel glide reflections
    Orbit space – the Klein bottle

The shaded area in each picture is a so-called *fundamental region* for the group, that is to say its images under all the group elements fill out the entire plane, and if two such intersect they do so only in their boundaries. So no two points in the interior of a fundamental region are identified by a group element. Of course, a fundamental region can be chosen in many different ways and its shape is by no means unique.

These three groups are members of a family which can be described as follows. We consider the group of all isometries of the plane and we assume known the fact that an isometry can be written as an ordered pair $(\theta, \mathbf{v})$ where $\theta \in O(2)$ and $\mathbf{v} \in \mathbb{E}^2$. So θ is either a rotation about the origin or a reflection in some line through the origin, and $\mathbf{v}$ has the effect of a translation. The isometry acts on $\mathbb{E}^2$ by

$$(\theta, \mathbf{v})(\mathbf{x}) = \theta(\mathbf{x}) + \mathbf{v}$$

and group multiplication is given by

$$(\theta, \mathbf{v})(\phi, \mathbf{w}) = (\theta\phi, \theta(\mathbf{w}) + \mathbf{v})$$

We give this group of isometries the topology of the product space $O(2) \times \mathbb{E}^2$ and call the resulting topological group the *euclidean group* E(2). (Note that although E(2) has the product topology from the spaces O(2) and $\mathbb{E}^2$, the group structure is *not* the product structure. Algebraically we have in fact the *semidirect product* of O(2) and $\mathbb{E}^2$.)

If G is a discrete subgroup of E(2), that is to say the topology induced from E(2) makes G into a discrete space, and if the orbit space $\mathbb{E}^2/G$ is compact, then G is called a *plane-crystallographic group*.

Our three examples rather clearly fit this description, and the orbit spaces are the torus, sphere, and Klein bottle respectively. (In each case take a fundamental region and work out what identifications have to be made to its sides in order to form $\mathbb{E}^2/G$.)

If G is a plane-crystallographic group, and if p is a point of the plane which is not left fixed by any non-identity element of G, then

$$\{x \in \mathbb{E}^2 \mid \|x - p\| \leqslant \|x - g(p)\| \text{ for all } g \in G\}$$

is a convex polygon which is a fundamental region for G, and G is a subgroup of finite index in the full group of symmetries of the resulting tessellation of the plane. The compactness of $\mathbb{E}^2/G$ ensures that this fundamental region is bounded.

Plane-crystallographic groups can be classified and they fall into precisely 17 distinct isomorphism classes.† Higher-dimensional crystallographic groups are defined in the same sort of way, and the number of isomorphism classes for a particular dimension is always finite.‡

Problems

26. Give an action of $\mathbb{Z}$ on $\mathbb{E}^1 \times [0,1]$ which has the Möbius strip as orbit space.

27. Find an action of $\mathbb{Z}_2$ on the torus with orbit space the cylinder.

28. Describe the orbits of the natural action of SO(n) on $\mathbb{E}^n$ as a group of linear transformations, and identify the orbit space.

29. If $\pi : X \to X/G$ is the natural identification map, and if O is open in X, show that $\pi^{-1}(\pi(O))$ is the union of the sets $g(O)$ where $g \in G$. Deduce that π takes open sets to open sets. Does π always take closed sets to closed sets?

30. Show that X may be Hausdorff yet X/G non-Hausdorff. If X is a compact topological group and G a closed subgroup acting on X by left translation, show that X/G is Hausdorff.

31. The stabilizer of a point $x \in X$ consists of those elements g in G for which $g(x) = x$. Show that the stabilizer of any point is a closed subgroup of G when X is Hausdorff, and that points in the same orbit have conjugate stabilizers for any X.

32. If G is compact, X Hausdorff, and if G acts transitively on X, show that X is homeomorphic to the orbit space $G/(\text{stabilizer of } x)$ for any $x \in X$.

---

† H. S. M. Coxeter, *Introduction to Geometry*, Wiley, 1961.

R. L. E. Schwarzenberger, 'The 17 Plane Symmetry Groups', *Mathematical Gazette*, 1974.

‡ One of Hilbert's problems, solved by Bieberbach (1911).

33. Let p, q be integers which have highest common factor 1. Let P be a regular polygonal region in the plane with centre of gravity at the origin and vertices $a_0, a_1, \ldots, a_{p-1}$, and let X be the solid double pyramid formed from P by joining each of its points by straight lines to the points $b_0 = (0,0,1)$ and $b_q = (0,0,-1)$ of $\mathbb{E}^3$ (see Fig. 4.6). Identify the triangles with vertices a_i, a_{i+1}, b_0, and a_{i+q},

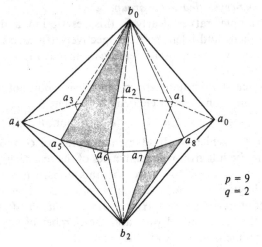

Figure 4.6

a_{i+q+1}, b_q for each $i = 0,1,\ldots, p-1$, in such a way that a_i is identified to a_{i+q}, a_{i+1} to a_{i+q+1}, and b_0 to b_q. (The subscripts $i+1$, $i+q$, $i+q+1$ are of course read mod p.) Prove that the resulting space is homeomorphic to the Lens space $L(p,q)$.

34. Show that $L(2,1)$ is homeomorphic to P^3. If p divides $q - q'$, prove that $L(p,q)$ is homeomorphic to $L(p,q')$.

5. The Fundamental Group

*On a dit souvent que la géométrie est l'art de bien
raisonner sur les figures mal faites.*

H. POINCARÉ

5.1 Homotopic maps

We gave a brief description of how to set about defining the fundamental group
of a space at the end of Chapter 1. We recall that the idea is to manufacture a
group out of the set of loops in the space which begin and end at some specified
point (usually referred to as the base point).

By a *loop* in a space X we shall understand a map $\alpha: I \to X$ such that
$\alpha(0) = \alpha(1)$,† and we shall say that the loop is *based* at the point $\alpha(0)$. If α and β
are two loops based at the same point of X, we define the *product* $\alpha.\beta$ to be the
loop given by the formula

$$\alpha.\beta(s) = \begin{cases} \alpha(2s) & 0 \leqslant s \leqslant \frac{1}{2} \\ \beta(2s - 1) & \frac{1}{2} \leqslant s \leqslant 1. \end{cases}$$

Notice that $\alpha.\beta$ is continuous, maps $[0,\frac{1}{2}]$ onto the image of α in X, and maps
$[\frac{1}{2},1]$ onto the image of β.

Unfortunately, this multiplication does not give a group structure on the set
of loops based at a particular point; it is a simple matter to check that it is not
even associative. To resolve this problem and obtain a group we agree to
identify two loops if one can be continuously deformed into the other, keeping
the base point fixed throughout the deformation. The object of this section is to
say exactly what we mean by a continuous deformation.

We shall work in a rather more general setting: if $f,g: X \to Y$ are maps, we
shall consider what it means to deform f continuously into g. Such a continuous deformation will be called a *homotopy*. Intuitively, we would like a
family $\{f_t\}$ of maps from X to Y, one for each point t of $[0,1]$, with $f_0 = f$,
$f_1 = g$, and the property that f_t changes in a continuous fashion as t varies
between 0 and 1. To capture this notion of continuous change we make use of
the product space $X \times I$, observing that a map $F: X \times I \to Y$ gives rise to a
family $\{f_t\}$ if we set $f_t(x) = F(x,t)$.

---

† A slight change from the terminology used in Chapter 1, where a loop was a map from a circle
  to X.

(5.1) Definition. *Let* f,g : X → Y *be maps. Then* f *is homotopic to* g *if there exists a map* F : X × I → Y *such that* F(x,0) = f(x) *and* F(x,1) = g(x) *for all points* x ∈ X.

The map F is called a homotopy from f to g and we shall write $f \underset{F}{\simeq} g$. If, in addition, f and g agree on some subset A of X, we may wish to deform f to g without altering the values of f on A. In this case we ask for a homotopy F from f to g with the additional property that

$$F(a,t) = f(a) \qquad \text{for all } a \in A, \text{ for all } t \in I.$$

When such a homotopy exists we say that f is homotopic to g *relative to* A and write $f \underset{F}{\simeq} g$ rel A.

If we have two loops $\alpha, \beta : I \to X$ based at the same point p of X, then asking that α can be continuously deformed into β without moving the base point p is exactly the same as asking that α be homotopic to β relative to the subset $\{0,1\}$ of I. A homotopy from α to β rel $\{0,1\}$ is by definition a map F from the square $I \times I$ to X which sends the bottom of the square via α, the top via β, and the two vertical sides to the base point p. This last condition means that the restriction of F to any horizontal line $I \times \{t\}$ in the square is a loop based at p: sliding the line from the bottom of the square to the top gives a continuous family of loops starting at α and finishing at β. Figure 5.1 illustrates this situation for two loops on a torus. Of course the picture is very much simplified: in reality the loops α and β may cross themselves (and one another) and the image of the square $I \times I$ in the torus may be extremely complicated.

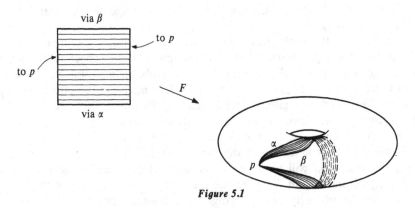

Figure 5.1

Examples of homotopies
1. Let C be a convex subset of a euclidean space and let $f,g : X \to C$ be maps, where X is an arbitrary topological space. For each point x of X, the straight line joining $f(x)$ to $g(x)$ lies in C, and we can define a homotopy from f to g simply by sliding f along these straight lines. To be precise, define $F : X \times I \to C$ by $F(x,t) = (1 - t)f(x) + tg(x)$. Notice that if f and g happen to agree on a

subset A of X then this homotopy is a homotopy relative to A. The homotopy F is called a *straight-line homotopy*.

2. Let $f,g : X \to S^n$ be maps which if evaluated on the same point of X never give a pair of antipodal points of S^n (i.e., $f(x)$ and $g(x)$ are never at opposite ends of a diameter). If we take S^n to be the unit sphere in $\mathbb{E}^{n+1}$, and think of f, g as maps into $\mathbb{E}^{n+1}$, then we have a straight-line homotopy from f to g. Since $f(x)$ and $g(x)$ are not antipodal, the straight line joining them does not pass through the origin. Therefore we can define $F : X \times I \to S^n$ by

$$F(x,t) = \frac{(1-t)f(x) + tg(x)}{\|(1-t)f(x) + tg(x)\|}$$

This map is a homotopy from f to g.

3. Let S^1 denote the unit circle in the complex plane, and consider the loops α, β in S^1 (both based at the point 1) defined by

$$\alpha(s) = \begin{cases} \exp 4\pi i s & 0 \leqslant s \leqslant \tfrac{1}{2} \\ \exp 4\pi i(2s - 1) & \tfrac{1}{2} \leqslant s \leqslant \tfrac{3}{4} \\ \exp 8\pi i(1 - s) & \tfrac{3}{4} \leqslant s \leqslant 1 \end{cases}$$

$$\beta(s) = \exp 2\pi i s \qquad 0 \leqslant s \leqslant 1.$$

Geometrically, α winds each of the segments $[0,\tfrac{1}{2}]$, $[\tfrac{1}{2},\tfrac{3}{4}]$, $[\tfrac{3}{4}, 1]$ once round the circle, the first two being wound in an anticlockwise direction, and the third clockwise. The loop β simply winds the whole interval $[0,1]$ once round the circle anticlockwise (Fig. 5.2).

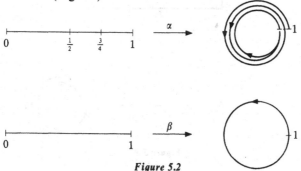

Figure 5.2

We can define a homotopy F from α to β relative to $\{0,1\}$ as follows, the continuity of our map being ensured by the glueing lemma (4.6):

$$F(s,t) = \begin{cases} \exp \dfrac{4\pi i s}{t+1} & 0 \leqslant s \leqslant \dfrac{t+1}{2} \\[2mm] \exp 4\pi i(2s - 1 - t) & \dfrac{t+1}{2} \leqslant s \leqslant \dfrac{t+3}{4} \\[2mm] \exp 8\pi i(1 - s) & \dfrac{t+3}{4} \leqslant s \leqslant 1. \end{cases}$$

This homotopy is illustrated in Fig. 5.3. We show the effect of F on the square $I \times I$, and the halfway stage $s \mapsto F(s,\frac{1}{2})$ of the homotopy.

(5.2) Lemma. *The relation of 'homotopy' is an equivalence relation on the set of all maps from* X *to* Y.

Proof. All maps f, g, h are from X to Y. For any f we have $f \underset{F}{\simeq} f$ where $F(x,t) = f(x)$, so the relation is reflexive. If $f \underset{F}{\simeq} g$ then $g \underset{G}{\simeq} f$ where $G(x,t) = F(x,1-t)$, giving symmetry. Finally, if $f \underset{F}{\simeq} g$ and $g \underset{G}{\simeq} h$, then $f \underset{H}{\simeq} h$ where H is defined by

$$H(x,t) = \begin{cases} F(x,2t) & 0 \leqslant t \leqslant \frac{1}{2} \\ G(x,2t-1) & \frac{1}{2} \leqslant t \leqslant 1, \end{cases}$$

so the relation is transitive.

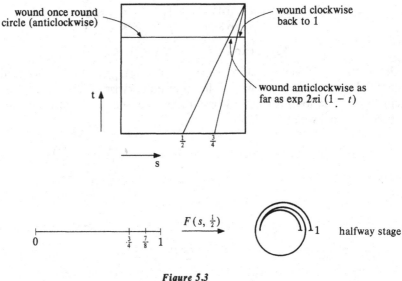

Figure 5.3

(5.3) Lemma. *The relation of 'homotopy relative to a subset* A *of* X' *is an equivalence relation on the set of all maps from* X *to* Y *which agree with some given map on* A.

Proof. If all the maps involved agree on A, then the homotopies defined above are homotopies relative to A.

(5.4) Lemma. *Homotopy behaves well with respect to composition of maps.*

Proof. We note that if we have maps

90

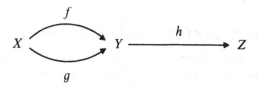

and if $f \underset{F}{\simeq} g$ rel A, then $hf \underset{hF}{\simeq} hg$ rel A as maps from X to Z.
Also given maps

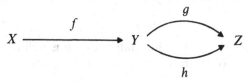

with $g \underset{G}{\simeq} h$ rel B for some subset B of Y, then $gf \underset{F}{\simeq} hf$ rel $f^{-1}B$ via the homotopy $F(x,t) = G(f(x),t)$.

Problems

1. Let C denote the unit circle in the plane. Suppose $f:C \to C$ is a map which is not homotopic to the identity. Prove that $f(x) = -x$ for some point x of C.

2. With C as above, show that the map which takes each point of C to the point diametrically opposite is homotopic to the identity. (We shall see later that the antipodal map of S^n is homotopic to the identity if and only if n is odd.)

3. Let D be the disc bounded by C, parametrize D using polar coordinates, and let $h:D \to D$ be the homeomorphism defined by $h(0) = 0$, $h(r,\theta) = (r, \theta + 2\pi r)$. Find a homotopy F from h to the identity map such that the functions $F|D \times \{t\}:D \times \{t\} \to D, 0 \leqslant t \leqslant 1$, are all homeomorphisms.

4. With the terminology of Problem 3, show that h is homotopic to the identity map relative to C.

5. Let $f:X \to S^n$ be a map which is not onto. Prove that f is *null homotopic*, that is to say f is homotopic to a map which takes all of X to a single point of S^n.

6. As usual, CY denotes the cone on Y. Show that any two maps $f,g:X \to CY$ are homotopic.

7. Show that a map from X to Y is null homotopic if and only if it extends to a map from the cone on X to Y.

8. Let A denote the annulus $\{(r,\theta) \mid 1 \leqslant r \leqslant 2, 0 \leqslant \theta \leqslant 2\pi\}$ in the plane, and let h be the homeomorphism of A defined by $h(r,\theta) = (r, \theta + 2\pi(r - 1))$. Show that h is homotopic to the identity map. Convince yourself that it is impossible to find a homotopy from h to the identity which is relative to the two boundary circles of A. (For a precise solution to this, see Problem 23.)

5.2 Construction of the fundamental group

Let X be a topological space, choose a base point $p \in X$, and consider the set of all loops in X based at p. As we have seen in Section 5.1 the relation of homotopy relative to $\{0,1\}$ is an equivalence relation on this set. We shall refer to the equivalence classes as *homotopy classes*, and denote the homotopy class of a loop α by $\langle \alpha \rangle$.

Multiplication of loops induces a multiplication of homotopy classes via

$$\langle \alpha \rangle . \langle \beta \rangle = \langle \alpha . \beta \rangle$$

Of course we must check that this multiplication is well defined. If $\alpha' \underset{F}{\simeq} \alpha$ rel $\{0,1\}$ and $\beta' \underset{G}{\simeq} \beta$ rel $\{0,1\}$, then $\alpha'.\beta' \underset{H}{\simeq} \alpha.\beta$ rel $\{0,1\}$, where

$$H(s,t) = \begin{cases} F(2s, t) & 0 \leqslant s \leqslant \frac{1}{2} \\ G(2s - 1, t) & \frac{1}{2} \leqslant s \leqslant 1 \end{cases}$$

(As usual, we refer to the glueing lemma to see that H is continuous.) Therefore $\langle \alpha' \rangle . \langle \beta' \rangle = \langle \alpha \rangle . \langle \beta \rangle$.

(5.5) Theorem. *The set of homotopy classes of loops in* X *based at* p *forms a group under the multiplication* $\langle \alpha \rangle . \langle \beta \rangle = \langle \alpha.\beta \rangle$.

Proof. We first check that multiplication is *associative*, i.e., $\langle \alpha.\beta \rangle . \langle \gamma \rangle = \langle \alpha \rangle . \langle \beta.\gamma \rangle$ for any three loops α, β, γ based at p. To do this we must show that $(\alpha.\beta).\gamma$ is homotopic to $\alpha.(\beta.\gamma)$ relative to $\{0,1\}$. One easily checks that $(\alpha.\beta).\gamma$ is equal to the composition $(\alpha.(\beta.\gamma)) \circ f$, where f is the map from I to I defined by

$$f(s) = \begin{cases} 2s & 0 \leqslant s \leqslant \frac{1}{4} \\ s + \frac{1}{4} & \frac{1}{4} \leqslant s \leqslant \frac{1}{2} \\ \dfrac{s + 1}{2} & \frac{1}{2} \leqslant s \leqslant 1 \end{cases}$$

Since I is convex and $f(0) = 0, f(1) = 1$, there is a straight-line homotopy from f to the identity map 1_I relative to $\{0,1\}$. By lemma (5.4) we have

$$(\alpha.\beta).\gamma = (\alpha.(\beta.\gamma)) \circ f$$

$$\simeq (\alpha.(\beta.\gamma)) \circ 1_I \text{ rel } \{0,1\}$$

$$= \alpha.(\beta.\gamma)$$

As usual, a diagram is much more effective than the formulae (Fig. 5.4).

The *identity element* is the homotopy class of the constant loop e at p defined by $e(s) = p$ for $0 \leqslant s \leqslant 1$. We can use a similar argument to the above to check that $\langle e \rangle . \langle \alpha \rangle = \langle \alpha \rangle$ and $\langle \alpha \rangle . \langle e \rangle = \langle \alpha \rangle$ for any loop α based at p. Consider the first of these. We need a homotopy relative to $\{0,1\}$ from $e.\alpha$ to α. Now $e.\alpha$ is the composition $\alpha \circ f$, where $f : I \to I$ is defined by

$$f(s) = \begin{cases} 0 & 0 \leqslant s \leqslant \frac{1}{2} \\ 2s - 1 & \frac{1}{2} \leqslant s \leqslant 1 \end{cases}$$

So

$$e.\alpha = \alpha \circ f \simeq \alpha \circ 1_I \text{ rel } \{0,1\}$$

$$= \alpha$$

We leave the verification of $\langle \alpha \rangle.\langle e \rangle = \langle \alpha \rangle$ to the reader.

Finally, we define the *inverse* of the homotopy class $\langle \alpha \rangle$ to be $\langle \alpha^{-1} \rangle$ where $\alpha^{-1}(s) = \alpha(1 - s), 0 \leqslant s \leqslant 1$. (So α^{-1} is just α 'in the opposite direction'.) The inverse is well defined, since if $\alpha \underset{F}{\simeq} \beta$ rel $\{0,1\}$ then $\alpha^{-1} \underset{G}{\simeq} \beta^{-1}$ where $G(s,t) = F(1 - s,t)$. To show $\langle \alpha \rangle.\langle \alpha^{-1} \rangle = \langle e \rangle$ we note that $\alpha.\alpha^{-1} = \alpha \circ f$ where $f : I \rightarrow I$ is defined by

$$f(s) = \begin{cases} 2s & 0 \leqslant s \leqslant \frac{1}{2} \\ 2 - 2s & \frac{1}{2} \leqslant s \leqslant 1. \end{cases}$$

Since $f(0) = f(1) = 0$, we know that $f \simeq g$ rel $\{0,1\}$, where $g(s) = 0, 0 \leqslant s \leqslant 1$. Therefore

$$\alpha.\alpha^{-1} = \alpha \circ f \simeq \alpha \circ g \text{ rel } \{0,1\}$$

$$= e$$

To show $\langle \alpha^{-1} \rangle.\langle \alpha \rangle = \langle e \rangle$ is no more difficult. This completes the proof of the theorem.

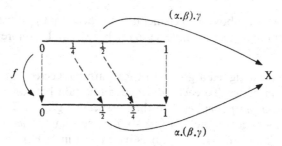

Figure 5.4

We have given a fairly painless proof of theorem (5.5) by leaning heavily on the fact that any two maps from the unit interval to itself which agree on 0 and 1 are homotopic relative to $\{0,1\}$. One can of course sit down and construct the necessary homotopies in a barehanded fashion (as in example 3 of Section 5.1) and we recommend the reader to do this for himself.

The group constructed in theorem (5.5) is called the *fundamental group of X based at* p, and written $\pi_1(X,p)$. Since any loop based at p must lie entirely inside the path component of X which contains p, we restrict ourselves to path-connected spaces. With this restriction, the fundamental group is independent (up to isomorphism) of the choice of base point, allowing us to refer to *the* fundamental group of a path connected space and use the notation $\pi_1(X)$.†

(5.6) Theorem. *If* X *is path-connected then* $\pi_1(X,p)$ *and* $\pi_1(X,q)$ *are isomorphic for any two points* p,q $\in$ X.

Before giving the proof, we observe that two *paths* γ, σ in a space which satisfy $\gamma(1) = \sigma(0)$ give rise to a new path $\gamma.\sigma$ via our product formula

$$\gamma.\sigma(s) = \begin{cases} \gamma(2s) & 0 \leqslant s \leqslant \tfrac{1}{2} \\ \sigma(2s - 1) & \tfrac{1}{2} \leqslant s \leqslant 1 \end{cases}$$

The following facts can be verified exactly as for loops.

(a) If $\gamma \simeq \gamma'$ rel $\{0,1\}$ and $\sigma \simeq \sigma'$ rel $\{0,1\}$ then $\gamma.\sigma \simeq \gamma'.\sigma'$ rel $\{0,1\}$.
(b) For any three paths γ, σ, δ satisfying $\gamma(1) = \sigma(0)$ and $\sigma(1) = \delta(0)$, we have $(\gamma.\sigma).\delta \simeq \gamma.(\sigma.\delta)$ rel $\{0,1\}$.
(c) If γ^{-1} is the path defined by $\gamma^{-1}(s) = \gamma(1 - s)$, then $\gamma.\gamma^{-1}$ is homotopic rel $\{0,1\}$ to the constant path at $\gamma(0)$; similarly $\gamma^{-1}.\gamma$ is homotopic to the constant path at $\gamma(1)$.

***Proof of theorem (5.6)*.** Choose a path γ which begins at p and ends at q (such a path exists because X is path-connected). If α is a loop based at p, then $(\gamma^{-1}.\alpha).\gamma$ is based at q and we define

$$\pi_1(X,p) \xrightarrow{\;\gamma_*\;} \pi_1(X,q)$$

$$\langle\alpha\rangle \longmapsto \langle\gamma^{-1}.\alpha.\gamma\rangle$$

Using (a), (b), and (c) above, it is elementary to check that γ_* is well defined, is a homomorphism, and has an inverse, namely $(\gamma^{-1})_*$. Therefore γ_* is an iso-morphism.

So far we have assigned a group to each path-connected topological space. We can do even better. To each continuous function between two spaces we can assign a *homomorphism* between their respective groups. The construction is very natural and geometric. Let $f : X \to Y$ be continuous, let p be the chosen base point in X, and choose $q = f(p)$ as base point in Y. For any loop α based at p in X, the composite function $f \circ \alpha$ is a loop based at q in Y; moreover, lemma (5.4) shows us that composing two homotopic loops with f gives loops which are homotopic in Y. Therefore we can define a function

$$f_* : \pi_1(X,p) \to \pi_1(Y,q)$$

by $f_*(\langle\alpha\rangle) = \langle f \circ \alpha\rangle$. Since $f \circ (\alpha.\beta) = (f \circ \alpha).(f \circ \beta)$, we see that f_* is a homo-morphism: we say that f_* is *induced* by f.

---

† π_1 because it is the first of a sequence of groups $\pi_1(X), \pi_2(X), \ldots$, the so-called homotopy groups of X.

Our construction immediately gives:

(5.7) Theorem. $(g \circ f)_* = g_* \circ f_*$ *whenever we have spaces and maps*
$X \xrightarrow{\ f\ } Y \xrightarrow{\ g\ } Z$.

We ought to be more careful: our statement of theorem (5.7) is really a convenient abbreviation. To be completely precise, we should make explicit mention of base points, i.e., choose base points $p \in X$, $q = f(p) \in Y$, $r = g(q) \in Z$, and say that $(g \circ f)_* : \pi_1(X,p) \to \pi_1(Z,r)$ is the composition

$$\pi_1(X,p) \xrightarrow{\ f_*\ } \pi_1(Y,q) \xrightarrow{\ g_*\ } \pi_1(Z,r)$$

In the special case where we have a homeomorphism $h : X \to Y$, we may apply theorem (5.7) to $X \xrightarrow{\ h\ } Y \xrightarrow{\ h^{-1}\ } X$, and to $Y \xrightarrow{\ h^{-1}\ } X \xrightarrow{\ h\ } Y$, obtaining

$$h^{-1}{}_* \circ h_* = (1_X)_* : \pi_1(X,p) \to \pi_1(X,p)$$

$$h_* \circ h^{-1}{}_* = (1_Y)_* : \pi_1(Y,h(p)) \to \pi_1(Y,h(p))$$

But an identity function quite clearly induces the identity homomorphism, and therefore $h_* : \pi_1(X,p) \to \pi_1(Y,h(p))$ is an isomorphism. *So homeomorphic (path-connected) spaces have isomorphic fundamental groups.*

We now have one way of attempting to distinguish between two path-connected topological spaces. We can try to compute their fundamental groups and then check whether or not these groups are isomorphic. If they are not isomorphic, the spaces are not homeomorphic. If the groups are isomorphic then we gain no information and we are left to look for a finer, more sophisticated invariant to distinguish between the spaces in question.

Problems

9. Let α, β, γ be loops in a space X, all based at the point p. Write out formulae for $(\alpha.\beta).\gamma$ and $\alpha.(\beta.\gamma)$, and work out a specific homotopy between these two loops. Make sure that your homotopy is a homotopy rel$\{0,1\}$.

10. Let γ, σ be two paths in the space X which begin at the point p and end at q. As in the proof of theorem (5.6), these paths induce isomorphisms γ_*, σ_* of $\pi_1(X,p)$ with $\pi_1(X,q)$. Show that σ_* is the composition of γ_* and the inner automorphism of $\pi_1(X,q)$ induced by the element $\langle \sigma^{-1} \gamma \rangle$.

11. Let X be a path-connected space. When is it true that for any two points $p,q \in X$ all paths from p to q induce the same isomorphism between $\pi_1(X,p)$ and $\pi_1(X,q)$?

12. Show that any indiscrete space has trivial fundamental group.

13. Let G be a path-connected topological group. Given two loops α, β based at e in G, define a map $F : [0,1] \times [0,1] \to G$ by $F(s,t) = \alpha(s).\beta(t)$, where the dot denotes multiplication in G. Draw a diagram to show the effect of this map on the square, and prove that the fundamental group of G is abelian.

14. Let $\mathbb{E}^3_+$ denote those points of $\mathbb{E}^3$ which have nonnegative final coordinate. Show that the space $\mathbb{E}^3_+ - \{(x,y,z) \mid y = 0, 0 \leqslant z \leqslant 1\}$ has trivial fundamental group.

5.3 Calculations

This section contains our first calculations. We shall deal with the circle and one or two other simple spaces: more general calculations will have to wait until Chapter 6.

Space	Fundamental group
Convex subset of $\mathbb{E}^n$	Trivial
Circle	$\mathbb{Z}$
$S^n, n \geqslant 2$	Trivial
$P^n, n \geqslant 2$	$\mathbb{Z}_2$
Torus	$\mathbb{Z} \times \mathbb{Z}$
Klein bottle	$\{a, b \mid a^2 = b^2\}$
Lens space $L(p, q)$	$\mathbb{Z}_p$

Convex subset of $\mathbb{E}^n$. In this case, we can shrink any loop to the constant loop at the base point by means of a straight-line homotopy. So the fundamental group of a convex subset of a euclidean space is the trivial group. A path-connected space whose fundamental group is trivial is said to be *simply connected*.

The circle. Identify the circle with the unit circle in the complex plane, and let $\pi : \mathbb{R} \to S^1$ denote the exponential mapping $x \mapsto e^{2\pi ix}$. All integers are identified to the point $1 \in S^1$ by the exponential map, and we choose this point as our base point.

Given an integer $n \in \mathbb{Z}$, let γ_n denote the path $\gamma_n(s) = ns$, $0 \leqslant s \leqslant 1$, joining 0 to n in $\mathbb{R}$. Then γ_n projects under π to a loop based at 1 in S^1. Also, $\pi \circ \gamma_n$ winds round the circle n times, in an anticlockwise direction for n positive, or clockwise if n is negative.

(5.8) Theorem. *The function* $\phi : \mathbb{Z} \to \pi_1(S^1, 1)$ *defined by* $\phi(n) = \langle \pi \circ \gamma_n \rangle$ *is an isomorphism.*

In order to prove theorem (5.8) we shall need the help of some lemmas. First note that if γ is any other path joining 0 to n in $\mathbb{R}$, then γ and γ_n are homotopic relative to $\{0,1\}$ and therefore project to homotopic loops in S^1.

(5.9) Lemma. ϕ *is a homomorphism.*

Proof. Given integers m, n, let σ be the path in $\mathbb{R}$ defined by $\sigma(s) = \gamma_n(s) + m$. Then $\pi \circ \sigma = \pi \circ \gamma_n$ and $\gamma_m.\sigma$ joins 0 to $m + n$. Therefore

$$\phi(m + n) = \langle \pi \circ \gamma_{m+n} \rangle = \langle \pi \circ (\gamma_m.\sigma) \rangle$$
$$= \langle (\pi \circ \gamma_m)(\pi \circ \sigma) \rangle = \langle (\pi \circ \gamma_m)(\pi \circ \gamma_n) \rangle$$
$$= \phi(m).\phi(n)$$

Our next job is to show that ϕ *is onto*. In order to do this we begin with an element of $\pi_1(S^1,1)$, represent the element by a loop α based at 1, and try to 'lift' this loop to a path γ in $\mathbb{R}$ which begins at 0. In other words, we try to find a path γ in $\mathbb{R}$ which satisfies $\pi \circ \gamma = \alpha$ and $\gamma(0) = 0$. Suppose we can do this, then the endpoint $\gamma(1)$ of γ projects to $\alpha(1) = 1$ in S^1, and therefore must be an integer n. By construction $\phi(n) = \langle \alpha \rangle$. This integer is called the *degree* of α, it measures the number of times α winds round the circle.

To carry out this lifting process we need to examine our identification map $\pi : \mathbb{R} \to S^1$ in more detail. Let U be the open set in S^1 formed by deleting the point -1, and consider the inverse image of U in $\mathbb{R}$. This is precisely the union of all open intervals of the form $(n - \frac{1}{2}, n + \frac{1}{2}), n \in \mathbb{Z}$. We note that these intervals are pairwise disjoint and that the restriction of π to any one of them is a *homeomorphism* of the interval with U. Similarly, if $V = S^1 - \{1\}$, the inverse image of V breaks up as a disjoint union of open sets in such a way that the restriction of π to any one of the open sets is a homeomorphism. Now $U \cup V$ is all of S^1. Therefore if we have a loop in S^1 we can try to break it up into segments so that each segment lies in either U or V, then lift these segments one by one back into $\mathbb{R}$ using the special properties of U and V noted above.

(5.10) Path-lifting lemma. *If* σ *is a path in* S^1 *which begins at the point 1, there is a unique path* $\tilde{\sigma}$ *in* $\mathbb{R}$ *which begins at 0 and satisfies* $\pi \circ \tilde{\sigma} = \sigma$.

Proof. The open sets $\sigma^{-1}(U)$, $\sigma^{-1}(V)$ give an open cover of $[0,1]$, so by Lebesgue's lemma (3.11) we know that we can find points $0 = t_0 < t_1 < \ldots < t_m = 1$ such that each $[t_i, t_{i+1}]$ lies in $\sigma^{-1}(U)$ or $\sigma^{-1}(V)$. We first define $\tilde{\sigma}$ on the subinterval $[0,t_1]$. Since σ begins at the point 1 we must have $\sigma([0,t_1]) \subseteq U$. Remember that $\pi \,|\, (-\frac{1}{2}, \frac{1}{2})$ is a homeomorphism from $(-\frac{1}{2}, \frac{1}{2})$ to U, and let f denote its inverse. Now set $\tilde{\sigma}(s) = f\sigma(s)$ for $0 \leqslant s \leqslant t_1$. Suppose, inductively, that we have defined $\tilde{\sigma}$ on $[0, t_k]$ and wish to extend our definition over $[t_k, t_{k+1}]$. If $\sigma([t_k, t_{k+1}]) \subseteq U$, and if $\tilde{\sigma}(t_k) \in (n - \frac{1}{2}, n + \frac{1}{2})$, we let g denote the inverse of $\pi \,|\, (n - \frac{1}{2}, n + \frac{1}{2})$ and set $\tilde{\sigma}(s) = g\sigma(s)$, $t_k \leqslant s \leqslant t_{k+1}$. If $\sigma([t_k, t_{k+1}]) \subseteq V$, then $\tilde{\sigma}(t_k) \in (n, n + 1)$ for some n. The restriction of π to $(n, n + 1)$ is a homeomorphism, with inverse, say h, and we can define $\tilde{\sigma}(s) = h\sigma(s)$, $t_k \leqslant s \leqslant t_{k+1}$. This completes our inductive definition of the lifted path $\tilde{\sigma}$. Notice that having defined $\tilde{\sigma}$ on $[0, t_k]$, there is only one way to extend it over $[t_k, t_{k+1}]$; therefore $\tilde{\sigma}$ is unique.

Of course, we could have stated lemma (5.10) in a more general form. If σ is a path in S^1 which begins at the point p, we can find a unique path $\tilde{\sigma}$ in $\mathbb{R}$ which satisfies $\pi \circ \tilde{\sigma} = \sigma$ and which begins at any preassigned point of $\pi^{-1}(p)$. Such a path $\tilde{\sigma}$ is called a *lift* of σ.

In order to show that ϕ is one–one we shall need to lift homotopies from the circle back into the real line. This can be done using the following result:

(5.11) Homotopy-lifting lemma. *If* $F: I \times I \to S^1$ *is a map such that* $F(0,t) = F(1,t) = 1$ *for* $0 \leqslant t \leqslant 1$, *there is a unique map* $\tilde{F}: I \times I \to \mathbb{R}$ *which satisfies*

$$\pi \circ \tilde{F} = F; \; and$$
$$\tilde{F}(0,t) = 0, \, 0 \leqslant t \leqslant 1.$$

Proof. We shall give only an outline, since the idea is precisely the same as for lemma (5.10). Subdivide $I \times I$ into squares by means of horizontal and vertical lines so that each square maps into U, or into V, under F. This requires an application of Lebesgue's lemma. We build up our definition of $\tilde{F}$ over these squares one at a time, beginning with the bottom row, working from left to right; then dealing with the second row in the same direction; etc. There is one point to be made: notice that when we want to extend the definition of $\tilde{F}$ over a particular square, the part of the square on which $\tilde{F}$ is already defined consists of either the left-hand edge, or the left-hand edge and the bottom. In both cases, this set is connected. This means that its image under $\tilde{F}$ lies entirely inside one of the components of $\pi^{-1}(U)$ or $\pi^{-1}(V)$ (according as F sends the square in question inside U or V), and we use the fact that the restriction of π to this component is a homeomorphism in order to complete the definition of $\tilde{F}$ over our square.

Proof of theorem (5.8). By lemmas (5.9) and (5.10), we know that $\phi: \mathbb{Z} \to \pi_1(S^1, 1)$ is a homomorphism and is onto. To see that ϕ is one–one, we argue as follows. Let $n \in \mathbb{Z}$ and suppose $\phi(n)$ is the identity element of $\pi_1(S^1, 1)$. This means that if we join 0 to n by a path γ, then $\pi \circ \gamma$ is a *null-homotopic* loop, i.e., is homotopic to the constant loop at the base point. Choose a specific homotopy F from the constant loop at $1 \in S^1$ to $\pi \circ \gamma$, and apply lemma (5.11) to find $\tilde{F}: I \times I \to \mathbb{R}$ which projects onto F and satisfies $\tilde{F}(0,t) = 0$, $0 \leqslant t \leqslant 1$.

Let P denote the union of the left- and right-hand edges and bottom of $I \times I$: then F maps all of P to 1. Since $\pi \circ \tilde{F} = F$, and since P is connected, we know that $\tilde{F}$ must map all of P to some integer. But $\tilde{F}$ sends the left-hand edge of $I \times I$ to 0, so $\tilde{F}(P) = 0$.

The path in $\mathbb{R}$ defined by $\tilde{F}(s, 1)$ is a lift of $\pi \circ \gamma$ which begins at 0, and must therefore be γ by the uniqueness part of lemma (5.10). Since $\tilde{F}(1,1) = 0$, we conclude that $\gamma(1) = n = 0$. Therefore the kernel of ϕ consists of the integer 0 alone and we have proved that ϕ is an isomorphism.

The n-sphere. To show that S^n has trivial fundamental group, for $n \geq 2$, we use the following result:

(5.12) Theorem. *Let X be a space which can be written as the union of two simply connected open sets U, V in such a way that U $\cap$ V is path-connected. Then X is simply connected.*

Proof. We show that any loop in X is homotopic to a product of loops each of which is contained in either U or V. This is enough to prove the theorem since U and V are both simply connected.

Choose a base point $p \in U \cap V$, and let $\alpha : I \to X$ be a loop based at p. Using Lebesgue's lemma (3.11), we can find points $0 = t_0 < t_1 < t_2 < \ldots < t_n = 1$ in I such that $\alpha([t_{k-1}, t_k])$ is always contained in U or V. Write α_k for the path $s \mapsto \alpha((t_k - t_{k-1})s + t_{k-1}), 0 \leq s \leq 1$. Join p to each point $\alpha(t_k), 1 \leq k \leq n-1$. by a path γ_k which lies in U if $\alpha(t_k) \in U$, *and* which lies in V if $\alpha(t_k) \in V$. If $\alpha(t_k) \in U \cap V$ we need to find γ_k in $U \cap V$; this poses no problem since we have assumed $U \cap V$ to be path-connected. Our loop α is homotopic to the product

$$(\alpha_1 . \gamma_1^{-1}) . (\gamma_1 . \alpha_2 . \gamma_2^{-1}) . (\gamma_2 . \alpha_3 . \gamma_3^{-1}) . \ldots . (\gamma_{n-1} . \alpha_n)$$

each member of which is a loop contained in U or V. Figure 5.5 illustrates the argument for the 2-sphere written as the union of two open discs.

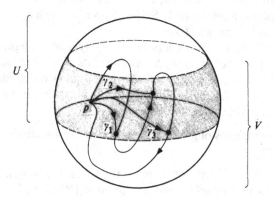

Figure 5.5

To apply this result to S^n, take distinct points x, y and set $U = S^n - \{x\}$, $V = S^n - \{y\}$. Both U and V are homeomorphic to $\mathbb{E}^n$, and therefore simply connected, and $U \cap V$ is path-connected provided $n \geq 2$.

Orbit spaces. The circle is the orbit space (Section 4.4) of the action of the integers on the real line by addition, and our computation of $\pi_1(S^1)$ is a special case of the following result, which also allows us to compute the fundamental groups of the torus, the Klein bottle, and the Lens space $L(p,q)$.

99

(5.13) Theorem. *If* G *acts as a group of homeomorphisms on a simply connected space* X, *and if each point* x ∈ X *has a neighbourhood* U *which satisfies* U ∩ g(U) = ∅ *for all* g ∈ G − {e},† *then* $\pi_1(X/G)$ *is isomorphic to* G.

Sketch proof. The idea is exactly as before. Fix a point $x_0 \in X$ and, given $g \in G$, join x_0 to $g(x_0)$ by a path γ. If $\pi: X \to X/G$ denotes the projection, $\pi \circ \gamma$ is a loop based at $\pi(x_0)$ in X/G. Define

$$\phi: G \to \pi_1(X/G, \pi(x_0))$$

by $\phi(g) = \langle \pi \circ \gamma \rangle$. Since X is simply connected, we can change γ to any other path joining x_0 to $g(x_0)$ without affecting ϕ.

It is not hard to check that ϕ is a homomorphism.‡ In order to prove ϕ one–one and onto we need analogues of our homotopy-lifting and path-lifting lemmas, (5.10) and (5.11). (For example, to show ϕ onto, start with an element $\langle \alpha \rangle \in \pi_1(X/G, \pi(x_0))$ and try to find a path γ in X which begins at x_0 and satisfies $\pi \circ \gamma = \alpha$. The endpoint $\gamma(1)$ lies in the orbit of x_0, so there is an element $g \in G$ such that $g(x_0) = \gamma(1)$. By construction $\phi(g) = \langle \alpha \rangle$.)

Thinking back to our work on the circle, we see that these two lemmas hold for any map $\pi: X \to Y$ with the following property.‡ For each $y \in Y$ we require an open neighbourhood V, and a decomposition of $\pi^{-1}(V)$ as a family $\{U_\alpha\}$ of pairwise disjoint open sets, in such a way that the restriction of π to each U_α is a homeomorphism from U_α to V. Such a map π is called a *covering map*, and X is called a *covering space*§ of Y.

Now given $y \in X/G$, we choose a point $x \in \pi^{-1}(y)$ and a neighbourhood U of x in X such that $U \cap g(U)$ is empty for all elements of G other than the identity. We set $V = \pi(U)$, remembering that $\pi: X \to X/G$ takes open sets to open sets, and take $\{g(U) \mid g \in G\}$ for the family $\{U_\alpha\}$. This shows that π is a covering map and completes our sketch proof of theorem (5.13).

Several of the examples of group actions in Section 4.4 satisfy the hypotheses of theorem (5.13):

Example 1. $\mathbb{Z} \times \mathbb{Z}$ on $\mathbb{E}^2$ with orbit space the torus T, giving $\pi_1(T) \cong \mathbb{Z} \times \mathbb{Z}$.
Example 2. $\mathbb{Z}_2$ on S^n with orbit space P^n, giving $\pi_1(P^n) \cong \mathbb{Z}_2$ for $n \geqslant 2$.
Example 6. $\mathbb{Z}_p$ on S^3 with orbit space the Lens space $L(p,q)$, giving $\pi_1(L(p,q)) \cong \mathbb{Z}_p$.

Consider example 1. Given a point of the plane, take the open disc of radius $\frac{1}{2}$ about this point as U. Then any translation in $\mathbb{Z} \times \mathbb{Z}$ (other than the identity) moves this disc off itself. We leave examples 2 and 6 to the reader.

Fundamental groups are *not* always abelian. Let G be the group with generators t, u, subject to the relation $u^{-1} tu = t^{-1}$, and consider the action of G on the plane determined by

---

† If this condition is satisfied then G has the discrete topology.
‡ Details are left to the reader, though we give some help in Problems 17–20.
§ For a detailed treatment of covering spaces see Chapter 10.

$$t(x,y) = (x + 1, y)$$

$$u(x,y) = (-x + 1, y + 1)$$

Then t is a translation parallel to the x axis, and u a glide reflection along the line $x = \frac{1}{2}$. The hypotheses of theorem (5.13) are easily checked, and the orbit space is the unit square with its sides identified as shown in Fig. 5.6, i.e., the Klein bottle K. Therefore the fundamental group of the Klein bottle is the group G. In terms of the parallel glide reflections $a = tu$, $b = u$, we recapture example 8c of Section 4.4 and we have $\pi_1(K) \cong \{a,b \mid a^2 = b^2\}$.

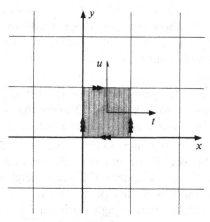

Figure 5.6

Product spaces. The final result of this section provides another tool for calculating fundamental groups:

(5.14) Theorem. *If* X *and* Y *are path-connected spaces* $\pi_1(X \times Y)$ *is isomorphic to* $\pi_1(X) \times \pi_1(Y)$.

Proof. Choose base points $x_0 \in X$, $y_0 \in Y$, and $(x_0, y_0) \in X \times Y$. All loops will be based at these points though, for simplicity, we shall omit them from the notation. The projections p_1, p_2 induce homomorphisms $p_{1*} : \pi_1(X \times Y) \rightarrow \pi_1(X)$, $p_{2*} : \pi_1(X \times Y) \rightarrow \pi_1(Y)$, and provide us with a ready-made homomorphism

$$\pi_1(X \times Y) \overset{\psi}{\longrightarrow} \pi_1(X) \times \pi_1(Y)$$

$$\langle \alpha \rangle \longmapsto (\langle p_1 \circ \alpha \rangle, \langle p_2 \circ \alpha \rangle)$$

If α is a loop in $X \times Y$, and if $p_1 \circ \alpha \underset{F}{\simeq} e_{x_0}$, $p_2 \circ \alpha \underset{G}{\simeq} e_{y_0}$, then $\alpha \underset{H}{\simeq} e_{(x_0, y_0)}$ where $H(s,t) = (F(s,t), G(s,t))$. Therefore ψ is one–one.

To show that ψ is onto, we begin with loops β in X, γ in Y, and form the loop $\alpha(s) = (\beta(s), \gamma(s))$ in $X \times Y$. By construction, $p_1 \circ \alpha = \beta$ and $p_2 \circ \alpha = \gamma$. Therefore $\psi(\langle \alpha \rangle) = (\langle \beta \rangle, \langle \gamma \rangle)$ as required.

101

This result gives a second proof that the torus has fundamental group $\mathbb{Z} \times \mathbb{Z}$ and shows, for example, that $\pi_1(S^m \times S^n)$ is the trivial group when $m,n \geqslant 2$.

Problems

15. Use theorem (5.13) to show that the Möbius strip and the cylinder both have fundamental group $\mathbb{Z}$.

16. Think of S^n as the unit sphere in $\mathbb{E}^{n+1}$. Given a loop α in S^n, find a loop β in $\mathbb{E}^{n+1}$ which is based at the same point as α, is made up of a finite number of straight line segments, and satisfies $\| \alpha(s) - \beta(s) \| < 1$ for $0 \leqslant s \leqslant 1$. Deduce from this that S^n is simply connected when $n \geqslant 2$. Where does your argument break down in the case $n = 1$?

17. Read through the sketch proof of theorem (5.13). If $g_1, g_2 \in G$, join x_0 to $g_1(x_0)$ by a path γ_1, and x_0 to $g_2(x_0)$ by γ_2. Observe that $\gamma_1.(g_1 \circ \gamma_2)$ joins x_0 to $g_1g_2(x_0)$ and deduce from this that ϕ is a homomorphism.

18. Let $\pi: X \to Y$ be a covering map. So each point $y \in Y$ has a neighbourhood V for which $\pi^{-1}(V)$ breaks up as a union of disjoint open sets, each of which maps homeomorphically onto V under π. Call such a neighbourhood 'canonical'. If α is a path in Y, show how to find points $0 = t_0 < t_1 < \ldots < t_m = 1$ such that $\alpha([t_i, t_{i+1}])$ lies in a canonical neighbourhood for $0 \leqslant i \leqslant m - 1$. Hence lift α piece by piece to a (unique) path in X which begins at any preassigned point of $\pi^{-1}(\alpha(0))$.

19. Let $\pi: X \to Y$ be a covering map, $p \in Y$, $q \in \pi^{-1}(p)$, and $F: I \times I \to Y$ a map such that $F(0,t) = F(1,t) = p$ for $0 \leqslant t \leqslant 1$. Use the argument of lemma (5.11) to find a map $\tilde{F}: I \times I \to X$ which satisfies $\pi \circ \tilde{F} = F$, and $\tilde{F}(0,t) = q, 0 \leqslant t \leqslant 1$. Check that $\tilde{F}$ is unique.

20. Redo Problem 19 as follows. For each t in $[0,1]$ we have a path $F_t(s) = F(s,t)$ in Y which begins at p. Let $\tilde{F}_t$ be its unique lift to a path in X which begins at q, and set $\tilde{F}(s,t) = \tilde{F}_t(s)$. Check that $\tilde{F}$ is continuous and lifts F.

21. Describe the homomorphism $f_*: \pi_1(S^1, 1) \to \pi_1(S^1, f(1))$ induced by each of the following maps:
(a) The antipodal map $f(e^{i\theta}) = e^{i(\theta + \pi)}, 0 \leqslant \theta \leqslant 2\pi$.
(b) $f(e^{i\theta}) = e^{in\theta}, 0 \leqslant \theta \leqslant 2\pi$, where $n \in \mathbb{Z}$.
(c) $f(e^{i\theta}) = \begin{cases} e^{i\theta} & 0 \leqslant \theta \leqslant \pi \\ e^{i(2\pi - \theta)}, & \pi \leqslant \theta \leqslant 2\pi \end{cases}$

22. In Section 4.4 we described three different actions of $\mathbb{Z}_2$ on the torus, and found the orbit spaces to be the sphere, the torus, and the Klein bottle. For each of these actions, describe the homomorphism from the fundamental group of the torus to that of the orbit space induced by the natural identification map.

23. Provide a precise solution to the second part of Problem 8 as follows. Let α, β be the paths in A defined by $\alpha(s) = (s + 1,0)$ and $\beta(s) = h\alpha(s), 0 \leqslant s \leqslant 1$.

Show that if h is homotopic to the identity relative to the two boundary circles of A, then the loop $\alpha^{-1}\beta$ is homotopic rel$\{0,1\}$ to the constant loop at the point $(1,0)$. Now check that this loop represents a nontrivial element of the fundamental group of A.

5.4 Homotopy type

The fundamental group is in fact left invariant by a much larger class of maps than the class of homeomorphisms. Like the other algebraic invariants which we shall construct later (homology groups and the Euler characteristic), it is an invariant of the so-called 'homotopy type' of a space.

(5.15) Definition. *Two spaces* X *and* Y *have the same homotopy type, or are homotopy equivalent, if there exist maps*

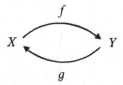

such that $g \circ f \simeq 1_X$ *and* $f \circ g \simeq 1_Y$.

The map g is called a homotopy inverse for f, and a map which has a homotopy inverse will be called a *homotopy equivalence*. We shall write $X \simeq Y$ when X and Y have the same homotopy type.

(5.16) Lemma. *The relation* X $\simeq$ Y *is an equivalence relation on topological spaces.*

Proof. The reflexive and symmetry properties are obvious. The relation is transitive because if we have maps

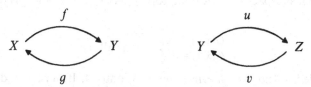

which are homotopy equivalences, then by lemma (5.4)

$$g \circ v \circ u \circ f \simeq g \circ 1_Y \circ f = g \circ f \simeq 1_X$$

and

$$u \circ f \circ g \circ v \simeq u \circ 1_Y \circ v = u \circ v \simeq 1_Z$$

Therefore the maps

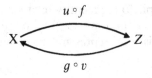

show that X and Z have the same homotopy type.

Examples.

1. Homeomorphic spaces have the same homotopy type.

2. Any convex subset of a euclidean space is homotopy equivalent to a point.

3. $\mathbb{E}^n - \{0\}$ has the homotopy type of S^{n-1}. Define $g:\mathbb{E}^n - \{0\} \to S^{n-1}$ by $g(\mathbf{x}) = \mathbf{x}/\|\mathbf{x}\|$, and let $f:S^{n-1} \to \mathbb{E}^n - \{0\}$ be inclusion. Then $g \circ f = 1_{S^n}$ and $1_{\mathbb{E}^n - \{0\}} \simeq f \circ g$ via $G(\mathbf{x}, t) = (1 - t)\mathbf{x} + t(\mathbf{x}/\|\mathbf{x}\|)$. The case $n = 2$ is illustrated by Fig. 5.7; the arrows indicate how points move during the homotopy G.

Figure 5.7

4. Let A be a subspace of X. A homotopy $G:X \times I \to X$ which is relative to A and for which

$$\left.\begin{array}{c} G(x,0) = x \\ G(x,1) \in A \end{array}\right\} \quad \text{for all } x \in X$$

will be called a *deformation retraction* of X onto A. If there is a deformation retraction of X onto A, then of course X and A have the same homotopy type (take $f:A \to X$ to be inclusion and $g:X \to A$ to be $x \mapsto G(x,1)$). Fig. 5.8 shows deformation retractions of a disc with two holes onto the one-point union of two circles (figure of eight); onto two circles joined by a line segment; and onto a space which looks like the letter θ. We conclude that all these spaces are homotopy equivalent. (Their fundamental group is the free group $\mathbb{Z} * \mathbb{Z}$ on two generators, as we shall see in Chapter 6.)

104

Suppose that $f,g : X \to Y$ are homotopic maps. As a first step towards showing that spaces of the same homotopy type have isomorphic fundamental groups, we propose to examine the relation between the homomorphisms f_*, g_* of fundamental groups induced by f and g. As we shall see they differ by an isomorphism.

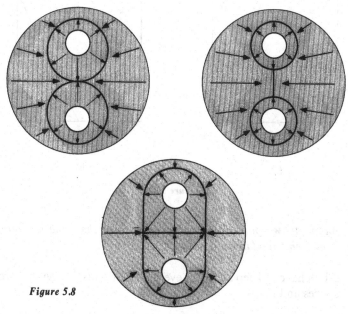

Figure 5.8

(5.17) Theorem. *If* $f \underset{\overline{F}}{\simeq} g : X \to Y$ *then* $g_* : \pi_1(X,p) \to \pi_1(Y,g(p))$ *is equal to the composition*

$$\pi_1(X,p) \xrightarrow{\ f_*\ } \pi_1(Y,f(p)) \xrightarrow{\ \gamma_*\ } \pi_1(Y,g(p)),$$

where γ *is the path joining* $f(p)$ *to* $g(p)$ *in* Y *defined by* $\gamma(s) = F(p,s)$.

Proof. Let α be a loop in X based at p. By definition, $g_*(\langle\alpha\rangle) = \langle g \circ \alpha\rangle$ and $\gamma_* f_*(\langle\alpha\rangle) = \langle \gamma^{-1}.(f\circ\alpha).\gamma\rangle$. We must therefore show that the loops $g\circ\alpha$ and $(\gamma^{-1}.(f\circ\alpha)).\gamma$ are homotopic relative to $\{0,1\}$.

Consider the map $G : I \times I \to Y$ defined by $G(s,t) = F(\alpha(s),t)$. This maps the sides of the square $I \times I$ as shown in Fig. 5.9a. Using G we construct a homotopy $H : I \times I \to Y$ between our two loops, whose effect on the square is illustrated in Fig. 5.9b, and whose precise definition is as follows:

$$H(s,t) = \begin{cases} \gamma(1 - 4s) & 0 \leqslant s \leqslant \dfrac{1-t}{4} \\[2mm] G\!\left(\dfrac{4s+t-1}{3t+1}, t\right) & \dfrac{1-t}{4} \leqslant s \leqslant \dfrac{1+t}{2} \\[2mm] \gamma(2s-1) & \dfrac{1+t}{2} \leqslant s \leqslant 1 \end{cases}$$

105

As usual, we appeal to the glueing lemma (4.6) to see that both G and H are continuous.

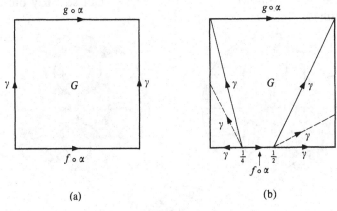

(a)   (b)

Figure 5.9

(5.18) Theorem. *If two path-connected spaces are of the same homotopy type, they have isomorphic fundamental groups.*

Proof. We shall have to keep a careful eye on base points during this proof. We are given spaces and maps

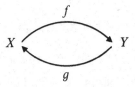

such that $1_X \underset{F}{\simeq} g \circ f$ and $1_Y \underset{G}{\simeq} f \circ g$. Choose a base point $p \in X$ which lies in the image of g, say $p = g(q)$. We shall show that $f_* : \pi_1(X,p) \to \pi_1(Y,f(p))$ is an isomorphism.

Let γ be the path joining p to $gf(p)$ in X defined by $\gamma(s) = F(p,s)$. Theorem 5.17 gives

$$(g \circ f)_* = \gamma_* : \pi_1(X,p) \longrightarrow \pi_1(X,gf(p))$$

which means that $(g \circ f)_*$ is an isomorphism. But $(g \circ f)_*$ is the composition

$$\pi_1(X,p) \xrightarrow{f_*} \pi_1(Y,f(p)) \xrightarrow{g_*} \pi_1(X,gf(p))$$

and therefore f_* is one–one.

To show f_* is onto, we proceed in a similar fashion. Let σ be the path joining q to $f(p)$ in Y defined by $\sigma(s) = G(q,s)$. By theorem (5.17)

$$(f \circ g)_* = \sigma_* : \pi_1(Y,q) \longrightarrow \pi_1(Y,f(p))$$

106

and so $(f \circ g)_*$ is an isomorphism. But $(f \circ g)_*$ is the composition

$$\pi_1(Y,q) \xrightarrow{g_*} \pi_1(X,p) \xrightarrow{f_*} \pi_1(Y,f(p))$$

and we see that f_* is onto. Therefore f_* is an isomorphism.

Using the above, we can squeeze a little more information out of our calculations. The Möbius strip, the cylinder, the punctured plane $\mathbb{E}^2 - \{0\}$, and the solid torus, all have the homotopy type of a circle, and consequently have $\mathbb{Z}$ as fundamental group. $\mathbb{E}^n - \{0\}$ deformation-retracts onto S^{n-1}, and is therefore a simply connected space when $n \geqslant 3$.

A space X is called *contractible* if the identity map 1_X is homotopic to the constant map at some point of X.

(5.19) Theorem. (a) *A space is contractible if and only if it has the homotopy type of a point.*
(b) *A contractible space is simply connected.*
(c) *Any two maps into a contractible space are homotopic.*
(d) *If X is contractible, then 1_X is homotopic to the constant map at x for any* $x \in X$.

Proof. (a) Given $p \in X$, write c_p for the constant map at p and i for the inclusion of $\{p\}$ in X. If 1_X is homotopic to c_p, the maps

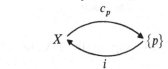

show X has the homotopy type of a point. Conversely, given maps

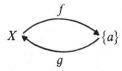

such that $g \circ f \simeq 1_X$, we see that 1_X is homotopic to the constant map at the point $p = g(a)$.
(b) If $1_X \simeq c_p$, the path $\gamma(s) = F(x,s)$ joins x to p. So X is path-connected.† Now apply theorem (5.18).
(c) If $1_X \simeq c_p$, then given maps $f,g : Z \to X$ we have

$$f = 1_X \circ f \simeq c_p \circ f = c_p \circ g \simeq 1_X \circ g = g$$

(d) Suppose $1_X \simeq c_p$ and apply (c) to the maps $c_p, c_x : X \to X$.

---

† If X and Y have the same homotopy type then X is path-connected if and only if Y is path-connected.

Any convex subset of a euclidean space is contractible, and we can easily imagine how to deform the identity map (along straight lines) to the constant map at any point. However, this example should not lead to too much optimism. In 'homotoping' the identity map 1_X to a constant map c_p, we may be forced to move the point p during the homotopy, i.e., there may not be a homotopy relative to $\{p\}$ from 1_X to c_p. For an example, take the 'comb space' shown in Fig. 5.10 as X, and take p to be the point $(0,\frac{1}{2})$. There is no homotopy from 1_X to c_p which keeps p fixed. (Why not?) But we can shrink each tooth of the comb vertically until we arrive at the interval $[0,1]$ on the x axis, then shrink this interval to the point 0:

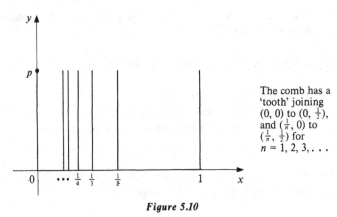

The comb has a 'tooth' joining $(0, 0)$ to $(0, \frac{1}{2})$, and $(\frac{1}{n}, 0)$ to $(\frac{1}{n}, \frac{1}{2})$ for $n = 1, 2, 3, \ldots$

Figure 5.10

this shows 1_X homotopic to c_0. Moving 0 up the y axis to p completes a homotopy from 1_X to c_p.

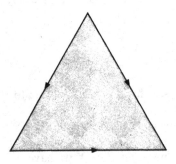

Figure 5.11

A contractible space may not look very contractible. If we identify the sides of a triangle in the manner indicated in Fig. 5.11 we obtain a space called the 'dunce hat'. The dunce hat is contractible (Problems 27, 28), though there appears no obvious way of setting about contracting it.

108

Problems

24. If $X \simeq Y$ and $X' \simeq Y'$, show that $X \times X' \simeq Y \times Y'$. Show also that CX is contractible for any space X.

25. Show that the punctured torus deformation-retracts onto the one-point union of two circles.

26. Consider the following examples of a circle C embedded in a surface S:
(a) $S = $ Möbius strip, $C = $ boundary circle;
(b) $S = $ torus, $C = $ diagonal circle
$$= \{(x,y) \in S^1 \times S^1 \mid x = y\};$$
(c) $S = $ cylinder, $C = $ one of boundary circles.
In each case, choose a base point in C, describe generators for the fundamental groups of C and S, and write down in terms of these generators the homomorphism of fundamental groups induced by the inclusion of C in S.

27. Prove that if $f, g : S^1 \to X$ are homotopic maps, then the spaces formed from X by attaching a disc using f and using g are homotopy equivalent; in other words, $X \cup_f D \simeq X \cup_g D$.

28. Use Problem 27, and the third example of a homotopy given in Section 5.1, to show that the 'dunce hat' has the homotopy type of a disc, and is therefore contractible.

29. Show that the 'house with two rooms' pictured in Fig. 5.12 is contractible.

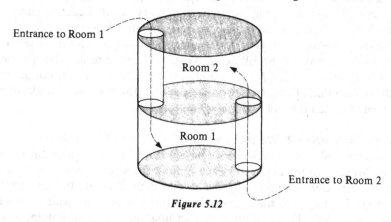

Entrance to Room 1

Room 2

Room 1

Entrance to Room 2

Figure 5.12

30. Give detailed proofs to show that the cylinder and Möbius strip both have the homotopy type of the circle.

31. Let X be the comb space shown in Fig. 5.10. Prove that the identity map of X is not homotopic rel$\{p\}$ to the constant map at p.

32. (*Fundamental theorem of algebra*) Show that any polynomial with complex coefficients, which is not constant, has a root in $\mathbb{C}$ as follows. We can clearly take the leading coefficient to be 1, so let $p(z) = z^n + a_{n-1}z^{n-1} + \ldots + a_1 z + a_0$.

Under the assumption that $p(z)$ is never zero, define a map $f_t : S^1 \to S^1$ by $f_t(z) = p(tz)/|p(tz)|$ for each nonnegative real number t. Prove that any two of these maps are homotopic, note that f_0 is a constant map, and produce a contradiction by showing that for t large enough, f_t is homotopic to the function $g(z) = z^n$.

5.5 The Brouwer fixed-point theorem

Our first application of the machinery created so far is to a celebrated result of L. E. J. Brouwer concerning fixed points of continuous functions. *Brouwer's theorem states that a continuous function from a ball (of any dimension) to itself must leave at least one point fixed.* For reasons which will emerge shortly, we cannot deal with the result in this degree of generality here. We shall give proofs assuming the dimension of the ball to be no more than two, leaving the general case to Chapter 8, theorem (8.14).

Proof for dimension 1. We can replace any ball of dimension 1, up to homeomorphism, by the unit interval $I = [0,1]$. We must show that if $f : I \to I$ is continuous, there is a point $x \in I$ such that $f(x) = x$. If not, then $I = \{x \in I \,|\, f(x) < x\} \cup \{x \in I \,|\, f(x) > x\}$. Now $f(1) < 1$ and $f(0) > 0$, so that these sets are nonempty, and using the continuity of f it is easy to check they are both open. Since I is connected, we have a contradiction.

A slightly different version of this argument, and one which lends itself better to higher dimensions, is the following. Again assume the result false, and define $g : I \to \{0,1\}$ by $g(x) = 0$ if $f(x) > x$ and $g(x) = 1$ if $f(x) < x$. The continuity of g follows from that of f, and g is onto since $g(0) = 0$ and $g(1) = 1$. We have once more contradicted the fact that I is a connected space.

Proof for dimension 2. We take the unit disc D in the plane as our standard two-dimensional ball and assume we have a map $f : D \to D$ which has no fixed points. Mimicking the above, for each point x draw a line segment from $f(x)$ to x (the direction is important) and extend it until it hits the unit circle C (Fig. 5.13). Sending x to the intersection of this line segment with C defines a function $g : D \to C$. The continuity of f ensures that g is continuous, and by construction $g(x) = x$ for all points of C.

We feel very strongly that a function $g : D \to C$, which is the identity on C, will have to tear D and therefore cannot possibly be continuous. In dimension 1 we obtained our contradiction by comparing the connectedness of I with the fact that $\{0,1\}$ is not connected. Both D and C are connected spaces, so we cannot use the same argument here. However, D is simply connected, whereas C has fundamental group $\mathbb{Z}$, and the contradiction now comes by arguing that the induced homomorphism $g_* : \pi_1(D) \to \pi_1(C)$ must be onto.

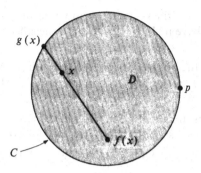

Figure 5.13

Take the point $p = (1,0)$ as base point for both C and D, and denote the inclusion of C in D by $i:C \to D$. The spaces and maps $C \xrightarrow{i} D \xrightarrow{g} C$ give rise to groups and homomorphisms

$$\pi_1(C,p) \xrightarrow{i_*} \pi_1(D,p) \xrightarrow{g_*} \pi_1(C,p)$$

Now $g \circ i(x) = x$ for all x in C, therefore $g_* \circ i_*$ is the identity homomorphism and g_* must be onto. But $\pi_1(D,p)$ is the trivial group and $\pi_1(C,p) \cong \mathbb{Z}$, so we have our contradiction and Brouwer's theorem must be true in dimension 2.

The above argument shows the interplay between algebra and topology at its best. The initial geometrical problem is difficult, yet once translated into algebra the solution uses only the simplest of ideas. Note the importance of theorem (5.7) in allowing us to identify the homomorphisms $g_* \circ i_*$ and $(g \circ i)_*$. For balls of dimension greater than 2 we can proceed in the same way, but we cannot use the fundamental group for the proof because the boundary of the n-ball (S^{n-1}) is simply connected for $n > 2$. We use homology groups instead; see Chapter 8.

If A is a subspace of X and if $g:X \to A$ is a map for which $g \mid A = 1_A$, then g is called a *retraction* of X onto A. With this terminology, the proof given above amounts to showing that there is no retraction of a disc onto its boundary circle. *The important property of a retraction is that it induces an onto homomorphism of fundamental groups.* (The proof is as above with D and C replaced by X and A respectively, and p replaced by a point of A.)

Problems

We shall say that the space X has the *fixed-point property* if every continuous function from X to itself has a fixed point.

33. Which of the following spaces have the fixed-point property?
(a) The 2-sphere; (b) the torus; (c) the interior of the unit disc; (d) the one-point union of two circles.

34. Suppose X and Y are of the same homotopy type and X has the fixed-point property. Does Y also have it? If X retracts onto the subspace A, and A has the fixed-point property, need X also have it?

35. Show that if X has the fixed-point property, and if X retracts onto the subspace A, then A also has the fixed-point property. Deduce the fixed-point property for the 'house with two rooms' of Problem 29.

36. Let f be a fixed-point-free map from a compact metric space to itself. Prove there is a positive number ε such that $d(x, f(x)) > \varepsilon$ for every point of the space.

37. Does the unit ball B^n in $\mathbb{E}^n$ with the point $(1,0,\dots,0)$ removed have the fixed-point property?

38. Show that the one-point union of X and Y has the fixed-point property if and only if both X and Y have it.

39. How does changing 'continuous function' to 'homeomorphism' in the definition of the fixed-point property affect Problems 33 and 37?

5.6 Separation of the plane

We say that a subset A of a space X *separates* X if $X - A$ has more than one component. In this section we shall prove two separation theorems for the plane:

(5.20) Theorem. *If* J *is a subspace of* $\mathbb{E}^2$ *which is homeomorphic to the circle, then* J *separates* $\mathbb{E}^2$.

(5.21) Theorem. *If* A *is a subspace of* $\mathbb{E}^2$ *which is homeomorphic to the closed interval* $[0,1]$, *then* A *does not separate* $\mathbb{E}^2$.

A subspace $J \subseteq \mathbb{E}^2$ homeomorphic to the circle is normally called a *Jordan curve*, or a *simple closed curve*. A subspace $A \subseteq \mathbb{E}^2$ homeomorphic to $[0,1]$ is called an *arc*. If $J \subseteq \mathbb{E}^2$ is a Jordan curve, then $\mathbb{E}^2 - J$ has (as one would expect) *exactly two* components, one bounded, the other unbounded, and J is the frontier of each. This is the famous Jordan curve theorem, a detailed discussion of which can be found in Munkres [10] and Wall [12]. We shall content ourselves here with the weaker statement of theorem (5.20), though we do give better results for polygonal curves in the problems.

Proof of theorem (5.20). We identify $\mathbb{E}^2$ with the plane in $\mathbb{E}^3$ determined by the equation $z = 0$, and we use S^2 to denote the unit sphere in $\mathbb{E}^3$. Let h be a homeomorphism from $\mathbb{E}^2$ to $S^2 - \{(0,0,1)\}$, choose a point $p \in h(J)$, and choose a homeomorphism $k : \mathbb{E}^2 \longrightarrow S^2 - \{p\}$.

Set $L = k^{-1}(h(J) - \{p\})$; then L is a closed subset of $\mathbb{E}^2$ which is homeomorphic to the real line. We imagine L as a line in the plane which runs off to infinity at both of its ends (Fig. 5.14). It is easy to check that $\mathbb{E}^2 - J, S^2 - h(J)$, and

$\mathbb{E}^2 - L$ all have the same number of components. We shall prove theorem (5.20) by showing† that $\mathbb{E}^2 - L$ *is not connected.*

We assume $\mathbb{E}^2 - L$ connected and aim for a contradiction. L is closed in $\mathbb{E}^2$ and therefore $\mathbb{E}^2 - L$ is path-connected by theorem (3.30). Let H_+, H_- denote the open half-spaces of $\mathbb{E}^3$ defined by $z > 0$, $z < 0$, and set

$$U = H_+ \cup \{(x,y,z) \mid (x,y) \in \mathbb{E}^2 - L, -1 < z \leqslant 0\}$$

$$V = H_- \cup \{(x,y,z) \mid (x,y) \in \mathbb{E}^2 - L, \quad 0 \leqslant z < 1\}$$

Then $U \cup V = \mathbb{E}^3 - L$, and $U \cap V$ is homeomorphic to $(\mathbb{E}^2 - L) \times (-1,1)$ which is a path-connected space. Also, both U and V are simply connected because any loop can be pushed vertically until it lies in either H_+ or H_-, and then shrunk to a point. Theorem (5.12) now tells us that $\mathbb{E}^3 - L$ is simply connected. To reach a contradiction, and hence to complete the proof of theorem (5.20), we use the following lemma.

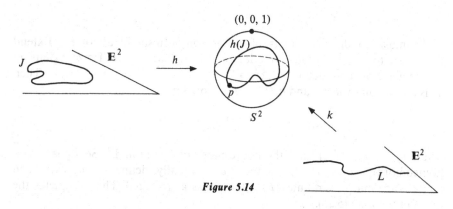

Figure 5.14

(5.22) Lemma. *There is a homeomorphism* $h : \mathbb{E}^3 \to \mathbb{E}^3$ *such that* $h(L)$ *is the z axis.*

If we can prove this lemma, then we have our contradiction as follows. By the lemma, $\mathbb{E}^3 - L$ is homeomorphic to $\mathbb{E}^3 - (z\ \text{axis})$, which is in turn homotopy equivalent to $\mathbb{E}^2 - \{0\}$. But the latter has infinite cyclic fundamental group. Therefore $\pi_1(\mathbb{E}^3 - L) \cong \mathbb{Z}$, contradicting the calculation made above.

Proof of (5.22). Choose a homeomorphism $f : L \to \mathbb{E}^1$ and consider the set of points $L_1 \subseteq \mathbb{E}^3$ defined by

$$L_1 = \{(x,y,f(x,y)) \mid (x,y) \in L\}$$

This is a closed line in $\mathbb{E}^3$ which lies vertically 'over' L and which intersects each horizontal plane in exactly one point. The idea is first to move L to L_1 by moving its points vertically, then to push L_1 horizontally across to the z axis (Fig. 5.15).

---

† Using an argument due to Doyle [24].

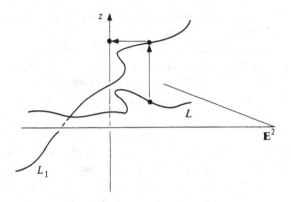

Figure 5.15

We must do this by means of a homeomorphism of all of $\mathbb{E}^3$. Extend $f:L \to \mathbb{E}^1$ to a continuous function $g:\mathbb{E}^2 \to \mathbb{E}^1$ using the Tietze extension theorem (2.15), and define $h_1:\mathbb{E}^3 \to \mathbb{E}^3$ by $h_1(x,y,z) = (x,y,z + g(x,y))$. Then h_1 is a homeomorphism and $h_1(L) = L_1$. Now set

$$h_2(x,y,z) = (x - f^{-1}(z)_x,\ y - f^{-1}(z)_y,\ z)$$

where $(f^{-1}(z)_x, f^{-1}(z)_y)$ are the coordinates of $f^{-1}(z)$ in $\mathbb{E}^2$. So h_2 is also a homeomorphism, and $h_2(L_1)$ is the z axis. Finally, define $h = h_2 \circ h_1$. Then h is a homeomorphism and $h(L)$ is the z axis as required. This completes the proof of theorem (5.20).

Proof of theorem (5.21). Suppose $\mathbb{E}^2 - A$ has more than one component. Since A is compact, and therefore bounded, $\mathbb{E}^2 - A$ has a unique unbounded component. Let K denote a bounded component of $\mathbb{E}^2 - A$. Choose a disc D with centre the origin and large enough so that $A \cup K$ lies in its interior. Let $p \in K$ and let $r:D - \{p\} \to S^1$ be the obvious retraction along straight lines joining p to the points of the boundary circle S^1 of D. Set $f = r\,|\,D - K:D - K \to S^1$.

Now consider $h = r\,|\,A:A \to S^1$. Since A is homeomorphic to $[0,1]$ we can use lemma (5.10) to lift h to a map $\bar{h}:A \to \mathbb{R}$ which satisfies $\pi \circ \bar{h} = h$, where $\pi:\mathbb{R} \to S^1$ is the exponential map. By the Tietze extension theorem, $\bar{h}$ extends to a map $\tilde{g}:A \cup K \to \mathbb{R}$. Set $g = \pi \circ \tilde{g}:A \cup K \to S^1$.

We plan to glue f and g together to give a map $f \cup g:D \to S^1$. Now f and g certainly agree on A; the only question is whether $f \cup g$ is continuous. The components of an open subset of a euclidean space are always open sets, so K is open. Therefore $D - K$ is closed in D. Also, the closure of K cannot meet any other component of $\mathbb{E}^2 - A$, so $\bar{K} \subseteq A \cup K$. Since A is clearly closed in D we see that $A \cup K$ is closed in D. By the glueing lemma, $f \cup g:D \to S^1$ is con-

114

tinuous. But $f \cup g(x) = f(x) = x$ for all points x of S^1, in other words, $f \cup g$ is a retraction. We have seen in Section 5.5 that there is no retraction of a disc onto its boundary, and we have the required contradiction.

Problems

40. Let A be a compact subset of $\mathbb{E}^n$. Show that $\mathbb{E}^n - A$ has exactly one unbounded component.

41. Let J be a polygonal Jordan curve in the plane. Choose a point p in the unbounded component of $\mathbb{E}^2 - J$ which does not lie on any of the lines produced by extending each of the segments of J in both directions. Given a point x of $\mathbb{E}^2 - J$, say that x is inside (outside) J if the straight line joining p to x cuts across J an odd (even) number of times. Show that the complement of J has exactly two components, namely the set of inside points and the set of outside points.

42. Let J be a polygonal Jordan curve in the plane, and let X denote the closure of the bounded component of J. Show that X can be broken up into a number of convex regions by extending the edges of J, then divide each of these regions into triangles. Now use induction on the number of triangles to show that X is homeomorphic to a disc.

43. Having done Problem 42, show there is a homeomorphism of the plane which takes J to the unit circle. (This is the Schönflies theorem for polygonal Jordan curves. It is true for a general Jordan curve, but much harder to prove.)

44. If J is a Jordan curve in the plane, use theorem (5.21) to show that the frontier of any component of $\mathbb{E}^2 - J$ is J.

45. Give an example of a subspace of the plane which has the homotopy type of a circle, which separates the plane into two components, but which is not the frontier of both of these components.

46. Give examples of simple closed curves on the torus, and on the projective plane, which separate, and which fail to separate.

47. Let X be a subspace of the plane which is homeomorphic to a disc. Generalize the argument of theorem (5.21) to show that X cannot separate the plane.

48. Suppose X is both connected and locally path-connected. Show that a map $f: X \to S^1$ lifts to a map $\tilde{f}: X \to \mathbb{R}$ (in other words $\tilde{f}$ followed by the exponential map is precisely f) if and only if the induced homomorphism $f_* : \pi_1(X) \to \pi_1(S^1)$ is the zero homomorphism.

5.7 The boundary of a surface

A surface is a Hausdorff space S in which each point has a neighbourhood homeomorphic either to $\mathbb{E}^2$, or to the closed half space $\mathbb{E}^2_+$ (Fig. 1.16). The

115

interior of S consists of the points of S which have a neighbourhood homeo-morphic to $\mathbb{E}^2$. Those points $x \in S$ for which there is a neighbourhood U, and a homeomorphism $f : \mathbb{E}^2_+ \to U$ such that $f(0) = x$, form the *boundary* of S.

These definitions satisfy our intuition as to what 'interior' and 'boundary' should mean for a surface. We must, however, check that a point cannot lie both in the interior and on the boundary.

(5.23) Theorem. *The interior and boundary of a surface are disjoint.*

Proof. We shall assume the result false and obtain a contradiction. Suppose x lies both on the boundary and in the interior of S. This means we can find neighbourhoods U, V of x in S, and homeomorphisms

$$f : \mathbb{E}^2_+ \to U$$

$$g : \mathbb{E}^2 \to V$$

such that $f(0) = g(0) = x$. Choose a half-disc $D_1 \subseteq \mathbb{E}^2_+$, centre the origin and small enough so that $f(D_1) \subseteq V$. Set $\phi = g^{-1}f : D_1 \to \mathbb{E}^2$.

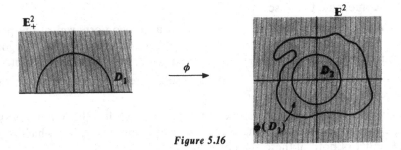

Figure 5.16

Because f and g are homeomorphisms, $\phi(D_1)$ must be a neighbourhood of 0 in $\mathbb{E}^2$. Choose a disc $D_2 \subseteq \mathbb{E}^2$ with centre the origin and of small enough radius so that $D_2 \subseteq \phi(D_1)$. Write ∂D_2 for the boundary circle of D_2 and let $r : \mathbb{E}^2 - \{0\} \to \partial D_2$ denote radial projection. Formally, if the radius of D_2 is R and if $y \in \mathbb{E}^2 - \{0\}$, then $r(y) = R(y/\|y\|)$. The restriction of r to $\phi(D_1) - \{0\}$ is a retraction of $\phi(D_1) - \{0\}$ onto ∂D_2, and should therefore induce a homomorphism of $\pi_1(\phi(D_1) - \{0\})$ *onto* $\pi_1(\partial D_2)$. But $\phi(D_1) - \{0\}$ is homeomorphic (via ϕ) to $D_1 - \{0\}$, and the latter is easily seen to be contractible. Therefore $\pi_1(\phi(D_1) - \{0\})$ is the trivial group, whereas $\pi_1(\partial D_2)$ is an infinite cyclic group. This gives us our contradiction.

(5.24) Theorem. *Let* $h : S_1 \to S_2$ *be a homeomorphism between two surfaces. Then* h *takes the interior of* S_1 *to the interior of* S_2, *and the boundary of* S_1 *to the boundary of* S_2.

Proof. If x lies in the interior of S_1, we can find a neighbourhood U of x in S_1 together with a homeomorphism $f : \mathbb{E}^2 \to U$. Since h is a homeomorphism, $h(U)$ is a neighbourhood of $h(x)$ in S_2, and $hf : \mathbb{E}^2 \to h(U)$ is a homeomorphism. Therefore $h(x)$ lies in the interior of S_2 and we have proved that h sends the interior of S_1 to that of S_2. The same argument can be applied to h^{-1}, so h maps the interior of S_1 *onto* the interior of S_2. Since the interior and boundary of a surface are disjoint, the boundary of S_1 must go onto the boundary of S_2 under h, completing the proof.

(5.25) Corollary. *Homeomorphic surfaces have homeomorphic boundaries.*

(5.26) Corollary. *The cylinder and the Möbius strip are not homeomorphic to one another.*

Problems

49. Use an argument similar to that of theorem (5.23) to prove that $\mathbb{E}^2$ and $\mathbb{E}^3$ are not homeomorphic.

50. Use the material of this section to show that the spaces X, Y illustrated in Problem 24 of Chapter 1 are not homeomorphic.

6. Triangulations

6.1 Triangulating spaces

The collection of all topological spaces is much too vast for us to work with. We have seen in previous chapters how to develop an abstract theory of topological spaces and continuous functions and to prove many important results. However, working in such a general setting we quickly run into two kinds of difficulty. On the one hand, in trying to prove a concrete geometrical result such as the classification theorem for surfaces, the purely topological structure of the surface (that it be locally euclidean) does not give us much leverage from which to start. On the other hand, although we can define algebraic invariants, such as the fundamental group, for topological spaces in general, they are not a great deal of use to us unless we can *calculate* them for a reasonably large collection of spaces. Both of these problems may be dealt with effectively by working with spaces that can be broken up into pieces which we can recognize, and which fit together nicely, the so called *triangulable* spaces.

Fig. 6.1 shows the sort of construction we have in mind. A homeomorphism from the surface of a tetrahedron to the sphere gives a decomposition of the sphere into four triangles, the triangles being joined along their edges. As a

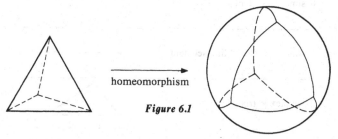

homeomorphism

Figure 6.1

second example, suppose we chop up a strip into triangles and then identify its ends with a half twist (Fig. 6.2). We obtain a space homeomorphic to a Möbius strip and we say that we have 'triangulated' the Möbius strip.

Both the sphere and the Möbius strip are surfaces. They are two-dimensional and so we can make models of them using triangles. For spaces of higher dimension we need higher-dimensional building blocks for our construction.

Let $v_0, v_1, \ldots, v_k$ be points of euclidean n-space $\mathbb{E}^n$. The hyperplane spanned by these points consists of all linear combinations $\lambda_0 v_0 + \lambda_1 v_1 + \ldots + \lambda_k v_k$, where each λ_i is a real number and the sum of the λ_i is 1. The points are in *general position* if any subset of them spans a strictly smaller hyperplane. It is an easy matter to check that if we regard $\mathbb{E}^n$ as a vector space, then this is equivalent to asking that the vectors $\mathbf{v}_1 - \mathbf{v}_0, \mathbf{v}_2 - \mathbf{v}_0, \ldots, \mathbf{v}_k - \mathbf{v}_0$ be linearly independent.

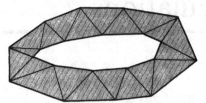

A simplicial complex homeomorphic
to the Möbius strip

Figure 6.2

Given $k + 1$ points $v_0, v_1, \ldots, v_k$ in general position, we call the smallest convex set containing them a *simplex of dimension k* (or a *k-simplex*). The points $v_0, v_1, \ldots, v_k$ are called the *vertices* of the simplex. We recall that a point x lies in the smallest convex set containing $v_0, v_1, \ldots, v_k$ if and only if it can be written as a linear combination

$$x = \lambda_0 v_0 + \lambda_1 v_1 + \ldots + \lambda_k v_k$$

where the λ_i are all nonnegative real numbers and $\lambda_0 + \lambda_1 + \ldots + \lambda_k = 1$. Looking at the first few dimensions we obtain:

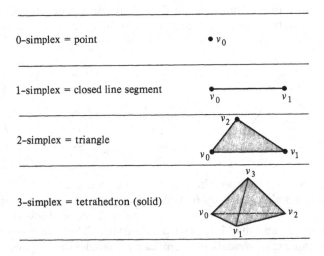

Simplexes have 'faces' in a natural way. If A and B are simplexes and if the vertices of B form a subset of the vertices of A, then we say that B is a *face* of A and write $B < A$. The idea of simplexes fitting together 'in a nice way' can be made precise by asking that if two simplexes intersect, then they do so in a common face (Fig. 6.3). We shall call a space *triangulable* if it is homeomorphic to the union of a finite collection of simplexes which fit together nicely in some euclidean space. We now look into this idea in a little more detail.

(6.1) Definition. *A finite collection of simplexes in some euclidean space* $\mathbb{E}^n$ *is called a simplicial complex if whenever a simplex lies in the collection then so does each of its faces, and whenever two simplexes of the collection intersect they do so in a common face.*

 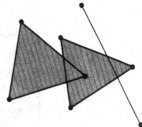

Simplexes which fit together nicely

The sort of intersections that are *not* allowed

Figure 6.3

We shall use letters such as K, L for simplicial complexes, reserving X and Y to denote topological spaces. Now the union of the simplexes which make up a particular complex† is a subset of a euclidean space, and can therefore be made into a topological space by giving it the subspace topology. A complex K, when regarded in this way as a topological space, is called a *polyhedron* and written $|K|$.

(6.2) Definition. *A triangulation of a topological space* X *consists of a simplicial complex* K *and a homeomorphism* $h: |K| \rightarrow X$.

Going back to our first example, X is the sphere, K the collection of simplexes which make up the surface of the tetrahedron, and, if the tetrahedron lies inside the sphere as in Fig. 1.8, h can be taken to be radial projection.

Asking that a space be triangulable is of course asking a great deal. A simplicial complex K is built up of a finite number of simplexes which live in a euclidean space, and consequently its polyhedron $|K|$ will have many pleasant properties: for example, it will be compact and a metric space. Therefore if a space is to be triangulable it must possess these properties. None the less, many important spaces admit a triangulation; in Chapter 7 we shall make essential use of the fact that all closed surfaces are triangulable.

---

† We often omit the word simplicial.

Triangulations are not unique.† The definition of a triangulation leaves us a great deal of choice, namely the choice of the simplicial complex K and of the triangulating homeomorphism h. A triangulation should be regarded as a *tool* which helps us to prove a particular result or do some calculation. It is its existence that is important: which triangulation we use is often of no great relevance.

A model for a triangulation of the torus is shown in Fig. 6.4. Making the identifications indicated via arrows on the edges of the rectangle, one can build a simplicial complex in $\mathbb{E}^3$ whose polyhedron is homeomorphic to the torus.

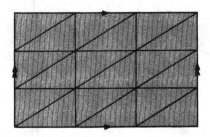

Figure 6.4

By definition, a simplicial complex always consists of simplexes which lie in some euclidean space $\mathbb{E}^n$. If we wish to emphasize the role played by the euclidean space, we say that K is a complex in $\mathbb{E}^n$. (We emphasize that K is a collection of simplexes and not a set of points.) Regard $\mathbb{E}^n$ as the subspace of $\mathbb{E}^{n+1}$ consisting of those points which have final coordinate zero. We can construct a complex CK in $\mathbb{E}^{n+1}$, which is called *the cone on K*, as follows. Let v denote the point $(0,0,\ldots,0,1)$ in $\mathbb{E}^{n+1}$. If A is a k-simplex in $\mathbb{E}^n$ with vertices $v_0, v_1, \ldots, v_k$, then the points $v_0, v_1, \ldots, v_k, v$ are in general position and therefore determine a $(k+1)$-simplex in $\mathbb{E}^{n+1}$. This $(k+1)$-simplex is called the *join* of A to v. Our cone CK consists of the simplexes of K, the join of each of these simplexes to v, and the 0-simplex v itself. One can easily check that the simplexes of this collection do fit together nicely and form a simplicial complex. CK is often called the join of K to v. As a set of points in $\mathbb{E}^{n+1}$, its polyhedron consists of all straight-line segments joining v to some point of $|K|$ (Fig. 6.5). In Chapter 4 we defined the cone CX on an arbitrary topological space X. The two ideas coincide in the sense that $|CK|$ and $C|K|$ are homeomorphic topological spaces (see lemma 4.5).

This cone construction gives us an easy way of triangulating the projective plane P. Recall that P is formed by taking a Möbius strip and a disc and sewing their boundaries together. Now we have already triangulated the Möbius strip M by means of the simplicial complex K in $\mathbb{E}^3$ shown in Fig. 6.2. Let L

---

† The only space with a unique triangulation consists of a single point.

consist of those simplexes of K which triangulate the boundary of M, i.e., the nineteen 1-simplexes and nineteen vertices which in our picture form the edge of K. Then $K \cup CL$ is a complex in $\mathbb{E}^4$ whose polyhedron is homeomorphic to the projective plane, for $|K|$ is homeomorphic to M and $|CL|$ is, up to homeomorphism, just a cone with base a circle, i.e., a disc. L as defined above is an

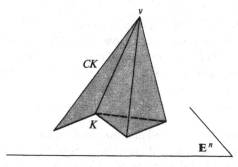

Figure 6.5

example of a *subcomplex* of a simplicial complex, i.e., it is a subcollection of the simplexes of a complex K which itself forms a complex.

In defining the cone on a complex K, we had to make a choice of point to represent the apex of the cone. We chose a point outside $\mathbb{E}^n$ to ensure that adding this new point to the set of vertices of a simplex of K produced a set of points in general position. But why choose v; why not choose some other point of $\mathbb{E}^{n+1} - \mathbb{E}^n$? A different choice would give a different set of simplexes in $\mathbb{E}^{n+1}$, but the simplexes would intersect one another in the same sort of way as the simplexes of CK. This leads us naturally to the idea of two simplicial complexes being isomorphic. Let K and L be complexes, not necessarily in the same euclidean space. They are *isomorphic* if there is a bijection ϕ from the set of vertices of K to the set of vertices of L such that $v_1, v_2, \ldots, v_s$ form the vertices of a simplex of K if and only if $\phi v_1, \phi v_2, \ldots, \phi v_s$ form the vertices of a simplex of L. The notion of isomorphism has nothing to do with the particular euclidean spaces in which the complexes lie, or the way in which their polyhedra are embedded in these euclidean spaces. It is simply a statement that K and L have the same number of simplexes of each dimension *and* that these simplexes exhibit the same pattern of intersections. The most important thing about isomorphic complexes is that they have *homeomorphic polyhedra*. Try to prove this. (The function ϕ is defined only on the vertices of K; try to extend it 'linearly' over each simplex of K to construct a homeomorphism from $|K|$ to $|L|$. We shall give the details of this construction in Section 6.3.) Now if $v,w \in \mathbb{E}^{n+1} - \mathbb{E}^n$, then the join of K to v and the join of K to w are isomorphic complexes (use the identity function on the vertices of K and send v to w). So our choice of apex in $\mathbb{E}^{n+1} - \mathbb{E}^n$ does not really matter.

We close this section by noting, for future reference, one or two facts concerning simplicial complexes. Let A be a simplex in $\mathbb{E}^n$ with vertices $v_0, v_1, \ldots, v_k$.

We define the *interior* of A to consist of those points x of A which can be written in the form $x = \lambda_0 v_0 + \lambda_1 v_1 + \ldots + \lambda_k v_k$ where $\sum_0^k \lambda_i = 1$ and the λ_i are *all positive*. Note that this notion coincides with the topological definition of interior when $k = n$, but not otherwise.

(6.3) Lemma. *Let* K *be a simplicial complex in* $\mathbb{E}^n$.
(a) $|K|$ *is a closed bounded subset of* $\mathbb{E}^n$, *and so* $|K|$ *is a compact space.*
(b) *Each point of* $|K|$ *lies in the interior of exactly one simplex of* K.
(c) *If we take the simplexes of* K *separately and give their union the identification topology, then we obtain exactly* $|K|$.
(d) *If* $|K|$ *is a connected space, then it is path-connected.*

Proof. Each simplex of K is closed and bounded. Since K is finite, the result (a) follows. For (b), suppose A and B are simplexes of K whose interiors overlap. Since K is a complex, A and B are required to meet in a common face. But the only face of a simplex which contains interior points is the whole simplex itself. Therefore $A = B$. In (c) we note that simplexes of K are closed subsets of $|K|$ since they are closed in $\mathbb{E}^n$. So if C is a subset of $|K|$, and if $C \cap A$ is closed in A for each simplex A of K, then $C \cap A$ must be closed in $|K|$. Therefore the finite union $C = \bigcup \{C \cap A \mid A \in K\}$ is closed in $|K|$. So the closed subsets of $|K|$ are precisely those which intersect each simplex of K in a closed set, in other words $|K|$ has the identification topology. Finally, for part (d), suppose $|K|$ is connected. Given $x \in |K|$, let L denote the subcomplex of K consisting of all those simplexes of K that do not contain x, and let ε denote the distance from x to $|L|$. Then if $\delta < \varepsilon$ the set $B(x,\delta) \cap |K|$ is path-connected, because any point in this set can be joined to x by a straight line in some simplex of K. This means that $|K|$ is a locally path-connected space, and we can mimic the proof of theorem (3.30) to show that it is path-connected.

Problems

1. Construct triangulations for the cylinder, the Klein bottle, and the double torus.

2. Finish off the proof of lemma (6.3).

3. If $|K|$ is a connected space, show that any two vertices of K can be connected by a path whose image is a collection of vertices and edges of K.

4. Check that $|CK|$ and $C|K|$ are homeomorphic spaces.

5. If X and Y are triangulable spaces, show that $X \times Y$ is triangulable.

6. If K and L are complexes in $\mathbb{E}^n$, show that $|K| \cap |L|$ is a polyhedron.

7. Show that S^n and P^n are both triangulable.

8. Show that the 'dunce hat' (Fig. 5.11) is triangulable, but that the 'comb space' (Fig. 5.10) is not.

6.2 Barycentric subdivision

Let K be a simplicial complex in $\mathbb{E}^n$. In this section we describe a construction which allows us to chop up the simplexes of K and produce a new complex K^1, which has the *same polyhedron* as K, but which has *simplexes of smaller diameter*.

The process is called 'barycentric subdivision'. If A is a simplex of K with vertices $v_0, v_1, \ldots, v_k$, then each point x of A has a unique expression of the form

$$x = \lambda_0 v_0 + \lambda_1 v_1 + \ldots + \lambda_k v_k \text{ where } \sum_0^k \lambda_i = 1 \text{ and all the } \lambda_i \text{ are nonnegative.}$$

These numbers λ_i are called the *barycentric coordinates* of the point x, and the *barycentre* (or centre of gravity) of A is the point

$$\hat{A} = \frac{1}{k+1}(v_0 + v_1 + \ldots + v_k).$$

In order to form K^1 we begin by adding extra vertices to K at the barycentres of its simplexes. Then, working in order of increasing dimension, we chop up each simplex of K as a cone with apex the extra vertex at its barycentre. Figure 6.6 illustrates the process.

To define K^1 precisely, we need to describe its simplexes. The vertices of K^1 are the barycentres of the simplexes of K. (This includes the original vertices of K since a 0-simplex is its own barycentre.)

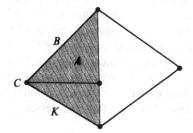 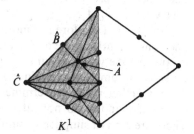

Figure 6.6

A collection $\hat{A}_0, \hat{A}_1, \ldots, \hat{A}_k$ of such barycentres form the vertices of a k-simplex of K^1 if and only if

$$A_{\sigma(0)} < A_{\sigma(1)} < \ldots < A_{\sigma(k)}$$

for some permutation σ of the integers $0, 1, 2, \ldots, k$. For example, in our illustration the barycentres $\hat{A}, \hat{B}, \hat{C}$ determine a 2-simplex of K^1, and looking at K we see $C < B < A$. Note that if $A_{\sigma(0)} < A_{\sigma(1)} < \ldots < A_{\sigma(k)}$, then for each i the barycentre $\hat{A}_{\sigma(i)}$ lies off the hyperplane spanned by $\hat{A}_{\sigma(0)}, \ldots, A_{\sigma(i-1)}$. Consequently, the points $\hat{A}_{\sigma(0)}, \ldots, \hat{A}_{\sigma(k)}$ are in general position.

The *dimension* of a simplicial complex K is the maximum of the dimensions of its simplexes, and its *mesh* $\mu(K)$ is the maximum of the diameters of its simplexes.

(6.4) Lemma. *The collection of simplexes described above forms a simplicial complex. It is denoted by* $\mathbf{K}^1$ *and is called the first barycentric subdivision of* $\mathbf{K}$. $\mathbf{K}^1$ *has the following properties:*
(a) *each simplex of* $\mathbf{K}^1$ *is contained in a simplex of* $\mathbf{K}$;
(b) $|\mathbf{K}^1| = |\mathbf{K}|$;
(c) *if the dimension of* $\mathbf{K}$ *is* n, *then* $\mu(\mathbf{K}^1) \leqslant \dfrac{n}{n+1}\mu(\mathbf{K})$.

Proof. If σ is a simplex of K^1, we can label its vertices $\hat{A}_0, \hat{A}_1, \ldots, \hat{A}_k$, where the A_i belong to K and $A_0 < A_1 < \ldots < A_k$. So all the vertices of σ lie in A_k, and therefore σ is contained in A_k. This proves property (a). Note that any face of σ lies in K^1, so in checking that K^1 is a simplicial complex we need only verify that its simplexes fit together nicely.

We shall prove that K^1 is a complex and satisfies $|K^1| = |K|$ by induction on the number of simplexes of K. The induction begins trivially when K consists of a single vertex. Suppose the result is true for all complexes which have less than m simplexes, and let K be a complex which is made up of m simplexes. Choose a simplex A of maximum dimension in K, and form a new complex L by removing A from K. Then L has $m - 1$ simplexes and its polyhedron consists of $|K|$ with the interior of the simplex A deleted. By the inductive hypothesis, L^1 is a simplicial complex and $|L^1| = |L|$. We need to look at the simplexes of K^1 that do not lie in L^1. Let σ be such a simplex (σ not equal to $\hat{A}$) and label its vertices as $\hat{A}_0, \hat{A}_1, \ldots, \hat{A}_{k-1}, \hat{A}$ where $A_0 < A_1 < \ldots < A_{k-1} < A$. The vertices $\hat{A}_0, \hat{A}_1, \ldots, \hat{A}_{k-1}$ determine a face τ of σ which lies in L^1, and $\tau = \sigma \cap |L^1|$. Therefore if σ meets a simplex of L^1, it must do so in a face of τ, and consequently in one of its own faces. Let σ' be a second simplex of $K^1 - L^1$ (again, not the vertex $\hat{A}$) and define τ' as above. Then if τ and τ' intersect, they do so in a common face (since L^1 is a complex). In this case the vertices of $\tau \cap \tau'$ together with $\hat{A}$ determine a common face of σ and σ' which is exactly $\sigma \cap \sigma'$. If τ and τ' do not intersect, then σ and σ' intersect in the vertex $\hat{A}$. Therefore K^1 is a simplicial complex.

Each simplex of K^1 being contained in a simplex of K, we know that $|K^1| \subseteq |K|$; so we now prove the reverse inclusion. Let $x \in |K|$ and let A be the unique simplex of K which contains x in its interior. If $x = \hat{A}$, then certainly $x \in |K^1|$. If not, join $\hat{A}$ to x by a straight line and prolong the line until it meets a face of A. Call the intersection point y. Then $y \in |L| = |L^1|$, and so $y \in \tau$ for some simplex τ of L^1. The vertices of τ together with $\hat{A}$ determine a simplex of K^1 which contains x. Therefore $x \in |K^1|$ and we have proved $|K^1| = |K|$, which is property (b).

It remains to verify property (c). First observe that the diameter of a simplex is the length of its longest edge. Let σ be an edge of K^1 with vertices $\hat{A}$ and $\hat{B}$, say, where $B < A$. Then σ is contained in A, and if the dimension of A is k we have

$$\text{length } \sigma \leqslant \frac{k}{k+1}(\text{diameter } A) \leqslant \frac{n}{n+1}(\text{diameter } A) \leqslant \frac{n}{n+1}\mu(K)$$

Therefore $\mu(K^1) \leqslant \dfrac{n}{n+1} \mu(K)$.

Define the *m-th barycentric subdivision* K^m of K inductively by $K^m = (K^{m-1})^1$. Figure 6.7 shows K^2 when K consists of a 2-simplex plus all its faces. Property (c) of Lemma (6.4) tells us that, by taking m large enough, we can make the diameters of the simplexes of K^m as small as we like.

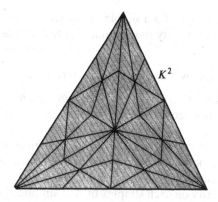

Figure 6.7

Problems

9. Make sure you can visualize the first barycentric subdivision of a 3-simplex.

10. Let $\mathscr{F}$ be an open cover of $|K|$. Show the existence of a barycentric subdivision K^r with the property that given a vertex v of K^r, there is an open set U in $\mathscr{F}$ which contains all the simplexes of K^r that have v as a vertex.

11. Let L be a subcomplex of K, and let N be the following collection of simplexes of K^2: a simplex B lies in N if we can find a simplex C in L^2 such that the vertices of B and C together determine a simplex of K^2. Show that N is a subcomplex of K^2, and that $|N|$ is a neighbourhood of $|L|$ in $|K|$.

12. Use the construction of Problem 11 to prove that if X is a triangulable space, and Y a subspace of X which is triangulated by a subcomplex of some triangulation of X, then the space obtained from X by shrinking Y to a point is triangulable.

6.3 Simplicial approximation

Let X and Y be topological spaces with triangulations $h:|K| \to X$, $k:|L| \to Y$. Then any map $f:X \to Y$ automatically induces a map $k^{-1}fh:|K| \to |L|$. There is a particular kind of map between polyhedra which is easy to work with, namely the so-called simplicial map which takes simplexes to simplexes, and which is *linear* on each simplex. In many problems, for example in calculating

127

the fundamental group of a triangulable space, it is important to be able to approximate a given map by a simplicial map. The approximation we choose will be close enough to the given map so that the two are homotopic; i.e., the approximation can be continuously deformed into the original map.

(6.5) Definition. *Let* K *and* L *be simplicial complexes. A function* $s: |K| \to |L|$ *is called simplicial if it takes simplexes of* K *linearly onto simplexes of* L.

Writing this out in detail: if A is a simplex of K, we require $s(A)$ to be a simplex of L; the condition of linearity means that if A has vertices $v_0, v_1, \ldots, v_k$, and if $x \in A$ is the point $x = \lambda_0 v_0 + \lambda_1 v_1 + \ldots + \lambda_k v_k$, where the λ_i are nonnegative and $\sum_0^k \lambda_i = 1$, then $s(x)$ when expressed in terms of the vertices of $s(A)$ is $s(x) = \lambda_0 s(v_0) + \lambda_1 s(v_1) + \ldots + \lambda_k s(v_k)$. Note that $s(A)$ may have lower dimension than A (we do not require s to be one–one), in which case $s(v_0), \ldots, s(v_k)$ will not all be distinct.

It should be clear that a simplicial function is *continuous*. This follows from the fact that a linear function between two simplexes is continuous, and application of the glueing lemma (4.6).

Because of its linearity on each simplex of K, a simplicial map s is completely determined once we know its effect on the vertices of K. In fact, if a function s from the vertices of K to the vertices of L has the property that if vertices $v_0, v_1, \ldots, v_k$ determine a simplex of K then $s(v_0), \ldots, s(v_k)$ determine a simplex of L, then s can be extended linearly across each simplex of K to give a simplicial map $|K| \to |L|$. In particular, an isomorphism from K to L extends in this way to a simplicial homeomorphism from the polyhedron of K to the polyhedron of L.

Now let $f: |K| \to |L|$ be a map between polyhedra. Given a point $x \in |K|$, the point $f(x)$ lies in the interior of a unique simplex of L. Call this simplex the *carrier* of $f(x)$.

(6.6) Definition. *A simplicial map* $s: |K| \to |L|$ *is a simplicial approximation of* $f: |K| \to |L|$ *if* $s(x)$ *lies in the carrier of* $f(x)$ *for each* $x \in |K|$.

Note that if s simplicially approximates f, then s and f are homotopic. This follows immediately from the definition. For suppose L lies in $\mathbb{E}^n$, and let $F: |K| \times I \to \mathbb{E}^n$ denote the straight-line homotopy defined by $F(x,t) = (1 - t)s(x) + tf(x)$. Given $x \in |K|$, we know that some simplex of L contains $s(x)$ and $f(x)$ and, since a simplex is convex, all points $(1 - t)s(x) + tf(x)$, $0 \le t \le 1$, must also lie in this simplex. Therefore the image of F lies in $|L|$, and F is a homotopy from s to f.

Simplicial approximations do not always exist (see Example (6.8) below). However, we can guarantee their existence if we are prepared to replace K by a suitable barycentric subdivision K^m.

(6.7) Simplicial approximation theorem. *Let* $f: |K| \to |L|$ *be a map between*

polyhedra. If m *is chosen large enough there is a simplicial approximation*
$s:|K^m| \to |L|$ *to* $f:|K^m| \to |L|$.

(6.8) Example. Let $|K| = |L| = [0,1]$, with K having vertices at the points
0, $\frac{1}{3}$, 1 and L at 0, $\frac{2}{3}$, 1 (Fig. 6.8). Suppose the given map $f:|K| \to |L|$ is

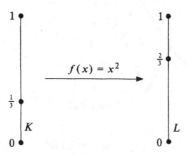

$$f(x) = x^2$$

Figure 6.8

$f(x) = x^2$. Then $f:|K| \to |L|$ does not admit a simplicial approximation. For if
$s:|K| \to |L|$ simplicially approximates f, then s must agree with f on the
inverse image of each vertex of L, so $s(0) = 0$ and $s(1) = 1$. But s is simplicial,
which forces $s(\frac{1}{3}) = \frac{2}{3}$. Therefore s takes the segment $[0, \frac{1}{3}]$ linearly onto $[0, \frac{2}{3}]$
and $[\frac{1}{3}, 1]$ linearly onto $[\frac{2}{3}, 1]$. We now have a contradiction, since the carrier of
$f(\frac{1}{2})$ is $[0, \frac{2}{3}]$ and this does not contain $s(\frac{1}{2})$. Similar reasoning shows there is no
simplicial approximation to $f:|K^1| \to |L|$. However, simplicial approxima-
tions to $f:|K^2| \to |L|$ do exist, and one such is shown in Fig. 6.9. We leave

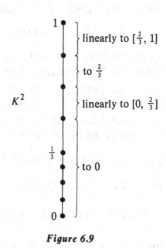

Figure 6.9

the reader to find a second, thereby showing that simplicial approximations
are *not unique*.

129

The proof of theorem (6.7) requires a lemma. Let K be a complex and let v be a vertex of K. The *open star* of v in K is the union of the interiors of those simplexes of K which have v as a vertex. It is an open subset of $|K|$ and we denote it by star (v,K) (Fig. 6.10).

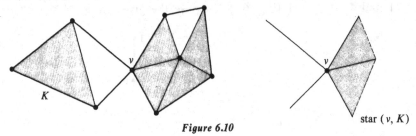

Figure 6.10

star (v, K)

(6.9) Lemma. *Vertices $v_0, v_1, \ldots, v_k$ of a simplicial complex K span (i.e., are the vertices of) a simplex of K if and only if the intersection of their open stars is nonempty.*

Proof. If $v_0, v_1, \ldots, v_k$ are the vertices of the simplex A of K then the whole of the interior of A lies in star(v_i, K) for $0 \leqslant i \leqslant k$. Conversely, suppose that $x \in \bigcap_0^k$ star (v_i, K) and let A be the carrier of x. By the definition of an open star, each v_i must be a vertex of A, and therefore $v_0, v_1, \ldots, v_k$ span some face of A.

Proof of theorem (6.7). We first deal with a special case of the theorem where it is not necessary to chop up the simplexes of K. Suppose that for each vertex u of K we can find a vertex v of L satisfying the inclusion

$$f(\text{star } (u,K)) \subseteq \text{star } (v,L) \qquad (*)$$

Define a function s from the vertices of K to those of L by choosing such a v for each u and setting $s(u) = v$. Then lemma (6.9) and the inclusion (*) give immediately that if $u_0, u_1, \ldots, u_k$ span a simplex of K, their images $s(u_0), s(u_1), \ldots, s(u_k)$ span a simplex of L. We can therefore extend s linearly over each simplex of K to give a simplicial map $s : |K| \to |L|$. This map s simplicially approximates f. For let x be a point of $|K|$ and let $u_0, u_1, \ldots, u_k$ be the vertices of its carrier. Then $x \in \bigcap_0^k$ star (u_i, K) and therefore by the inclusion (*) we have $f(x) \in \bigcap_0^k$ star $(s(u_i), L)$. This means that the carrier of $f(x)$ in L has the simplex spanned by $s(u_0), s(u_1), \ldots, s(u_k)$ as a face, and consequently it must contain the point $s(x)$.

To deal with the theorem in general, we need only show that we can arrange for inclusion (*) to be satisfied at the expense of replacing K by a suitable barycentric subdivision K^m. Now the open stars of the vertices of L form an open cover of $|L|$. Since $f : |K| \to |L|$ is continuous, the inverse images under f of these open sets give an open cover of $|K|$. Let δ be a Lebesgue number of this open cover ($|K|$ is a compact metric space so we can apply Lebesgue's lemma (3.11)) and choose m large enough so that $\mu(K^m) < \delta/2$. Given a vertex u of K^m, the diameter of its open star in K^m is less than δ, so star $(u,K^m) \subseteq f^{-1}$ (star (v,L)) for some vertex v of L, as required. This completes the proof.

The simplicial approximation theorem will be used in the next section in calculating fundamental groups, and again in Chapter 8 to check the topological invariance of the so-called homology groups of a space.

Problems

13. Use the simplicial approximation theorem to show that the n-sphere is simply connected for $n \geqslant 2$.

14. If $k < m,n$, show that any map from S^k to S^m is null homotopic, and that the same is true of any map from S^k to $S^m \times S^n$.

15. Show that a simplicial map from $|K|$ to $|L|$ induces a simplicial map from $|K^m|$ to $|L^m|$ for any m.

16. If $s:|K'''| \rightarrow |L|$ simplicially approximates $f:|K'''| \rightarrow |L|$, and $t:|L''| \rightarrow |M|$ simplicially approximates $g:|L''| \rightarrow |M|$, is $ts:|K'''^{+n}| \rightarrow |M|$ always a simplicial approximation for $gf:|K'''^{+n}| \rightarrow |M|$?

17. If $f:|K| \rightarrow |K|$ is a simplicial map, prove that the set of fixed points of f is the polyhedron of a subcomplex of K^1, though not necessarily of a subcomplex of K.

18. Use the simplicial approximation theorem to show that the set of homotopy classes of maps from one polyhedron to another is always countable.

19. Read the elegant proof of the Brouwer fixed-point theorem due to M. W. Hirsch given in Maunder [18].

6.4 The edge group of a complex

We calculated the fundamental groups of one or two spaces in Chapter 5, but our calculations, though efficient for the examples given there, were rather *ad hoc*. If we agree to work with triangulable spaces, we can be much more systematic. We shall show how to read off generators and relations† for the fundamental group from a triangulation of the space.

Let X be a path-connected triangulable space, take a specific triangulation $h:|K| \rightarrow X$, and replace X by $|K|$ (we are at liberty to do this since the fundamental group is a topological invariant). Now the advantage of a polyhedron

---

† The material on generators and relations, free groups and free products necessary for this section is collected together in the Appendix at the end of the book.

$|K|$ is that the elements of its fundamental group can be represented by loops which are made up of edges of K. Using such 'edge loops' we shall construct a group, called the edge group of the complex K, which can be computed and which is isomorphic to the fundamental group of $|K|$.

An edge path in a complex K is a sequence $v_0 v_1 \ldots v_k$ of vertices in which each consecutive pair $v_i v_{i+1}$ spans a simplex of K. For technical reasons, we allow the possibility $v_i = v_{i+1}$; if we apply a simplicial map to an edge path we want the result to be an edge path, even though two adjacent vertices may have been identified in the process. If $v_0 = v_k = v$, we have an *edge loop based at* v. In order to define the edge group of K, we need a simplicial version of the notion of homotopy. We consider two edge paths to be *equivalent* if we can obtain one from the other by a finite number of operations of the following type. If three vertices uvw span a simplex of K they may be replaced, in any edge path in which they occur consecutively, by the pair uw; under the same assumption the pair uw may be replaced by uvw. (Geometrically this allows us to replace two sides of a triangle by the third side, and vice versa, or to remove and introduce edges which make the path double back on itself; see Fig. 6.11.) In addition we allow ourselves to change a repeated vertex uu to u and vice versa.

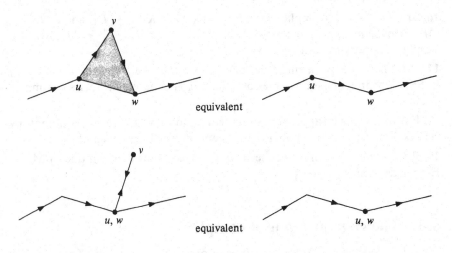

Figure 6.11

We shall denote the equivalence class of the edge path $v_0 v_1 \ldots v_k$ by $\{v_0 v_1 \ldots v_k\}$. One easily checks that the set of equivalence classes of edge loops based at a particular vertex v forms a group under the multiplication

$$\{vv_1 \ldots v_{k-1} v\} \cdot \{vw_1 \ldots w_{l-1} v\} = \{vv_1 \ldots v_{k-1} vw_1 \ldots w_{l-1} v\}$$

The identity element is the equivalence class $\{v\}$, and the inverse of $\{vv_1 \ldots v_{k-1} v\}$ is the class $\{vv_{k-1} \ldots v_1 v\}$. This is the *edge group of* K *based at* v; it will be written $E(K,v)$.

(6.10) Theorem. $E(K,v)$ *is isomorphic to* $\pi_1(|K|,v)$.

Proof. We construct a function $\phi : E(K,v) \to \pi_1(|K|,v)$ by simply interpreting each edge loop in K as a loop in $|K|$. Formally, given an edge loop $vv_1 \dots v_{k-1}v$, divide the unit interval I into k equal segments and let $\alpha : I \to |K|$ be the linear extension of

$$\alpha(0) = \alpha(1) = v, \qquad \alpha(i/k) = v_i \qquad 1 \leqslant i \leqslant k - 1$$

Then α is a loop in $|K|$ based at v. Since equivalent edge paths plainly give homotopic loops, we may define

$$\phi(\{vv_1 \dots v_{k-1}v\}) = \langle \alpha \rangle$$

It should be clear that ϕ is a homomorphism.

To show that ϕ is onto, we begin with a loop $\alpha : I \to |K|$ based at v, regard I as the polyhedron of a complex L which consists of the 1-simplex $[0,1]$ and its two vertices, and apply the simplicial approximation theorem to produce a simplicial map $s : |L^m| \to |K|$ which is homotopic to α. The vertices of L^m are the points $i/2^m$, $0 \leqslant i \leqslant 2^m$, and s picks out the edge loop $vv_1 \dots v_{2^m-1}v$ of K, where $v_i = s(i/2^m)$, $1 \leqslant i \leqslant 2^m - 1$. By construction,

$$\phi(\{vv_1 \dots v_{2^m-1}v\}) = \langle s \rangle = \langle \alpha \rangle$$

To complete our proof we must show that ϕ is one–one. Suppose $vv_1 \dots v_{k-1}v$ is an edge loop which, when interpreted as a loop in $|K|$ gives a null-homotopic loop α. We must prove that $vv_1 \dots v_{k-1}v$ is equivalent to the edge loop consisting of the single vertex v. Since α is null homotopic, we have a homotopy $F : I \times I \to |K|$ which satisfies

$$F(s,0) = \alpha(s), \qquad 0 \leqslant s \leqslant 1$$

and which sends the other three sides of the square to v. We think of $I \times I$ as the polyhedron of the complex L shown in Fig. 6.12, where a,b,c,d denote

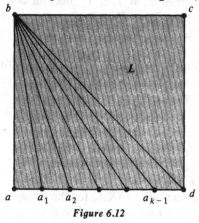

Figure 6.12

the four corners of the unit square and a_i stands for the point $(i/k, 0)$, and we note that $F(a_i) = v_i, 1 \leqslant i \leqslant k - 1$.

The two edge paths $a\, a_1 a_2 \ldots a_{k-1} d$ and $abcd$ are clearly equivalent in L. If we take a barycentric subdivision L^m of L, we obtain two edge paths in L^m which have $2^m - 1$ vertices inserted between each pair of vertices of the original paths. We shall denote these new paths by the symbols E_1, E_2, so as to avoid introducing notation for all the extra vertices. One can check by induction on m (Problem 21) that E_1 and E_2 are equivalent in L^m.

The simplicial approximation theorem gives us a barycentric subdivision L^m and a simplicial approximation $S:|L^m| \to |K|$ to $F:|L^m| \to |K|$. Now if two edge paths differ by a single operation of the type introduced in defining our notion of equivalence, and if we apply a simplicial map to the two paths, then their images will also differ by a single such operation. In other words, a simplicial map preserves the relation of equivalence between edge paths. Therefore the images under S of E_1 and E_2 are equivalent in K. But S applied to E_2 just gives the vertex v repeated a total of $3.2^m + 1$ times, which is equivalent to the edge loop v. Since $F(a_i) = v_i, 1 \leqslant i \leqslant k - 1$, and since S simplicially approximates F, the image under S of each new vertex in E_1 introduced between a_i and a_{i+1} is either v_i or v_{i+1}. Therefore applying S to E_1 gives an edge loop equivalent to $v\, v_1 v_2 \ldots v_{k-1}\, v$. This completes the argument.

We now turn to the problem of reading off generators and relations for $E(K,v)$. Let L be a subcomplex of K which contains all the vertices of K and for which $|L|$ is path-connected and simply connected. Such a subcomplex always exists: we can in fact build one using the edges of K as follows. A one-dimensional subcomplex of K whose polyhedron is both path-connected and simply connected is called a *tree*.

(6.11) Lemma. *A maximal tree contains all the vertices of* K.

Proof. Let T be a maximal tree in K; maximal means of course that if T' is a tree and contains T then $T' = T$. If T does not contain all the vertices of K, then some vertex v must lie in $K - T$. Choose a vertex u of T and, remembering that $|K|$ is path-connected, join u and v by a path in $|K|$. By the simplicial approximation theorem, we may replace this path by an edge path $u\, v_1 v_2 \ldots v_k\, v$. Let v_i be the last vertex of this edge path which lies in T, and form a new subcomplex T' by adding the vertex v_{i+1} and the edge spanned by $v_i\, v_{i+1}$ to T. The space $|T'|$ is just $|T|$ with a 'spike' attached, and it clearly deformation retracts onto $|T|$. Therefore T' is a tree, contradicting the maximality of T.

Suppose then that we have chosen our subcomplex L. Since $|L|$ is simply connected, edge loops in L will not contribute to $E(K,v)$, and therefore we can effectively ignore the simplexes of L in our calculations. List the vertices of K as $v = v_0, v_1, v_2, \ldots, v_s$, and write $G(K,L)$ for the group which is determined by generators g_{ij}, one for each ordered pair of vertices v_i, v_j that span a simplex of

K, subject to the relations $g_{ij} = 1$ if v_i, v_j span a simplex of L, and $g_{ij} g_{jk} = g_{ik}$ if v_i, v_j, v_k span a simplex of K.

(6.12) Theorem. $G(K,L)$ *is isomorphic to* $E(K,v)$.

The above description of $G(K,L)$ is designed to facilitate the proof of theorem (6.12). However, we can do a little better and rid ourselves of some unwanted generators. Notice that setting $i = j$ gives $g_{ii} = 1$, and setting $i = k$ shows $g_{ji} = g_{ij}^{-1}$. Therefore we need only introduce a generator g_{ij} for each pair of vertices v_i, v_j which span an edge of $K - L$ and for which $i < j$. The first type of relation is now redundant, and the only ones of the second type which matter are the relations $g_{ij} g_{jk} = g_{ik}$ whenever v_i, v_j, v_k† span a 2-simplex of $K - L$ and $i < j < k$.

Proof of theorem (6.12). We shall construct homomorphisms

$$G(K,L) \xrightarrow{\phi} \xleftarrow{\theta} E(K,v)$$

which are inverse to one another. Join v to each vertex v_i of K by an edge path E_i in L, taking $E_0 = v$, and define ϕ on the generators of $G(K,L)$ by

$$\phi(g_{ij}) = \{E_i v_i v_j E_j^{-1}\}$$

If v_i, v_j span a simplex of L, then $E_i v_i v_j E_j^{-1}$ is an edge loop which lies entirely in L, and therefore represents the identity element of $E(K,v)$ since $|L|$ is simply connected. Also, if v_i, v_j, v_k span a simplex of K, we have

$$\phi(g_{ij})\phi(g_{jk}) = \{E_i v_i v_j E_j^{-1}\}\{E_j v_j v_k E_k^{-1}\}$$
$$= \{E_i v_i v_j E_j^{-1} E_j v_j v_k E_k^{-1}\}$$
$$= \{E_i v_i v_j v_k E_k^{-1}\}$$
$$= \{E_i v_i v_k E_k^{-1}\}$$
$$= \phi(g_{ik})$$

So the relations in $G(K,L)$ are preserved and ϕ defines a homomorphism from $G(K,L)$ to $E(K,v)$.

It is not hard to check that the function

$$\theta(\{v\, v_k\, v_l\, v_m \ldots v_n\, v\}) = g_{0k}\, g_{kl}\, g_{lm} \ldots g_{n0}$$

defines a homomorphism from $E(K,v)$ to $G(K,L)$. Now

$$\theta\phi(g_{ij}) = \theta(\{E_i v_i v_j E_j^{-1}\}) = g_{ij},$$

---

† If two of these vertices, say v_i, v_j, span a simplex of L, we interpret g_{ij} as 1.

since the pairs of vertices in E_i and E_j^{-1} span simplexes of L. So $\theta\phi$ is the identity. Further, for any edge loop $v\,v_k\,v_l\ldots v_n\,v$ we have

$$\{v\,v_k\,v_l\ldots v_n\,v\} = \{E_0\,v\,v_k\,E_k^{-1}\}\{E_k\,v_k\,v_l\,E_l^{-1}\}\ldots\{E_n\,v_n\,v\,E_0^{-1}\}.$$

But $\phi\theta$ is the identity on each of the terms in this product, and therefore $\phi\theta$ is the identity homomorphism.

Examples

1. Take X to be the one-point union of n circles (often called a 'bouquet of circles'), and triangulate each circle using the boundary of a triangle, labelling the vertices as illustrated in Fig. 6.13 for the case $n = 3$. Let L consist of the two edges from each triangle which contain the common vertex v, plus all the vertices. Then $E(K,v) \cong G(K,L)$ is generated by n elements $g_{12}, g_{34}, \ldots, g_{2n-1, 2n}$, and there are no relations. So $\pi_1(X)$ is the *free group* on n generators. For the case of a single circle we obtain the free group $\mathbb{Z}$ on a single generator, agreeing with our earlier calculations.

Note that if X is path-connected and can be triangulated by means of a one-dimensional complex, then $\pi_1(X)$ is a free group since there are no 2-simplexes to enforce relations between the generators.

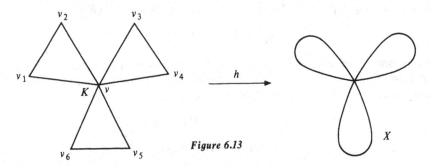

Figure 6.13

2. The fundamental group of any (path-connected) triangulable space is *finitely presented*, that is to say it is determined by a finite number of generators and a finite number of relations. (This follows because a complex is made up of a finite number of simplexes.)

3. The definition of $E(K,v)$ involves only the vertices, edges, and triangles of K. Therefore if $K(2)$ denotes the subcomplex of K consisting of those simplexes which have dimension at most 2, we have $\pi_1(|K|) \cong \pi_1(|K(2)|)$. This subcomplex $K(2)$ is called the 2-*skeleton* of K. Using this observation we can give a second, rather neat, proof that S^n is simply connected when $n \geqslant 2$. Triangulate S^n by the boundary of an $(n+1)$-simplex, and note that when $n \geqslant 2$ the 2-skeleton of the $(n+1)$-simplex and that of its boundary coincide. The result now follows since a simplex is contractible.

4. Triangulate the Klein bottle by means of the complex represented in Fig. 6.14, and let L be the shaded subcomplex. The 1-simplexes of $K - L$ provide us

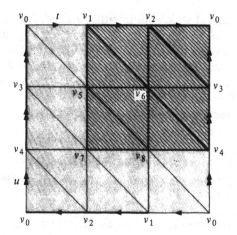

Figure 6.14

with eleven generators, and the 2-simplexes with ten relations between them. Set $t = g_{01}$ and $u = g_{04}$. The triangle spanned by the vertices v_0, v_1, v_5 gives

$$g_{01} g_{15} = g_{05}$$

in other words $t = g_{05}$, since the vertices v_1, v_5 span an edge of L. Working down the triangles in the left-hand column of Fig. 6.14, we obtain

$$t = g_{05} = g_{35} = g_{37} = g_{47}$$

$$g_{24} t = g_{27}$$

$$1 g_{24} = u.$$

Combine the last two relations to give $g_{27} = ut$. We now work along the remaining triangles in the bottom row, obtaining

$$ut = g_{17} = g_{18}$$
$$t u t = g_{08}$$
$$u 1 = g_{08}$$

So we are finally led to the single relation $t u t = u$. Therefore the fundamental group of the Klein bottle is given by two generators t, u subject to the single relation $t u t = u$. (It is worth comparing this with the calculation given in Chapter 5.)

This last example shows that, even for a very simple space, we may have so many generators and relations as to make practical calculation unpleasant. Luckily, we can use theorem (6.12) to produce a short cut. To this end, let J, K be simplicial complexes in the same euclidean space which intersect in a common subcomplex, and suppose that $|J|$, $|K|$, $|J \cap K|$ are all path-connected spaces. Imagine we know the fundamental groups of these three spaces and want to calculate $\pi_1(|J \cup K|)$.

Think first of the simplest possible case, namely when J and K intersect in a single vertex. Then any edge loop in $J \cup K$ based at this vertex is clearly a

137

product of loops, each of which lies in either J or K, and we expect to obtain the free product $\pi_1(|J|) * \pi_1(|K|)$ for the fundamental group of $|J \cup K|$. In the general case, the same sort of reasoning holds, except that the free product $\pi_1(|J|) * \pi_1(|K|)$ effectively counts the homotopy classes of those loops which lie in $|J \cap K|$ twice (once in each of $\pi_1(|J|)$, $\pi_1(|K|)$), and therefore we must correct this by adding some extra relations.

Let j, k denote the inclusion maps $|J \cap K| \subseteq |J|$, $|J \cap K| \subseteq |K|$, and take a vertex v of $J \cap K$ as base point.

(6.13) Van Kampen's theorem.† *The fundamental group of $|J \cup K|$ based at v is obtained from the free product $\pi_1(|J|, v) * \pi_1(|K|, v)$ by adding the relations $j_*(z) = k_*(z)$ for all‡ $z \in \pi_1(|J \cap K|, v)$.*

Proof. Take a maximal tree T_0 in $J \cap K$ and extend it to give maximal trees T_1, T_2 in J and K respectively. Then $T_1 \cup T_2$ is a maximal tree in $J \cup K$. By theorems (6.10) and (6.12), $\pi_1(|J \cup K|)$ is generated by elements g_{ij} corresponding to edges of $J \cup K - T_1 \cup T_2$, with relations $g_{ij} g_{jk} = g_{ik}$ given by the triangles of $J \cup K$. But this is precisely the group which results from taking a generator a_{ij} for each edge of $J - T_1$, a generator b_{ij} for each edge of $K - T_2$, with relations of the form $a_{ij} a_{jk} = a_{ik}$, $b_{ij} b_{jk} = b_{ik}$ corresponding to the triangles of J, K, and adding the extra relations $a_{ij} = b_{ij}$ whenever a_{ij} and b_{ij} correspond to the same edge of $J \cap K$. It remains only to note that the edges of $J \cap K - T_0$, when regarded as edges of J, give a set of generators for $j_*(\pi_1(|J \cap K|))$; similarly the same edges, when thought of as in K, generate $k_*(\pi_1(|J \cap K|))$.

Examples

1. We return to the triangulation of the Klein bottle given in Fig. 6.14, and let J be the complex which results from deleting the 2-simplex spanned by the vertices v_0, v_1, v_5. Then $|J|$ is the Klein bottle with an open disc punched out. For K we take the 2-simplex just mentioned together with all its faces. So $|K|$ is a disc and $|J \cap K|$ a circle.

Now the square with the interior of a triangle removed in this way deformation retracts onto its boundary. But in $|J|$ the edges of the square are identified so as to give two circles joined together at the point v_0, and the deformation retraction is compatible with these identifications since it leaves the boundary of the square fixed throughout. Therefore we have a deformation retraction of $|J|$ onto the one-point union of two circles, and we see that $\pi_1(|J|, v_0)$ is the free group $\mathbb{Z} * \mathbb{Z}$ with generators t, u represented by the sides of the square.

Choose a generator z for the infinite cyclic group $\pi_1(|J \cap K|, v_0)$ as in Fig. 6.15. Then $k_*(z)$ is the identity element, since $|K|$ is simply connected, and $j_*(z)$ is plainly identified by our deformation retraction with the word $t u^{-1} t u$ in $\pi_1(|J|, v_0)$. Van Kampen's theorem now tells us that $\pi_1(|J \cup K|, v_0)$ is

---

† Proved independently by H. Seifert and E. R. van Kampen.

‡ We need only add such relations for a set of generators of $\pi_1(|J \cap K|, v)$.

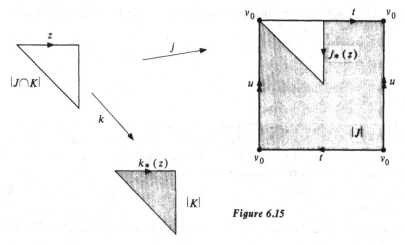

Figure 6.15

obtained from the group $(\mathbb{Z} * \mathbb{Z}) * \{e\}$ by adding the extra relation $t u^{-1} t u = e$. In other words, the Klein bottle has fundamental group

$$\{t,u \mid t u^{-1} t u = e\} = \{t,u \mid t u t = u\}$$

2. We often apply van Kampen's theorem without actually specifying triangulations for the spaces involved.† The triangulations were important as tools, enabling us to prove theorem (6.13), but the actual statement of the theorem is a statement about polyhedra (and therefore about triangulable spaces), and does not depend on them.

Suppose we think of the projective plane P as obtained by glueing together the boundary circles of a Möbius band and a disc. We know that the Möbius band has infinite cyclic fundamental group, and its boundary circle clearly represents twice a generator (Fig. 6.16). So van Kampen's theorem tells us that

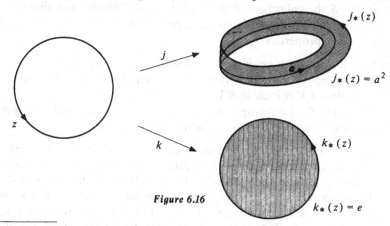

Figure 6.16

---

† We must, however, be sure that our spaces can be triangulated. Although there are more general versions of van Kampen's theorem (see for example Massey [9]), the result is false for arbitrary topological spaces.

139

$\pi_1(P)$ is obtained from the free product $\mathbb{Z} * \{e\}$ by adding the relation $a^2 = e$. In other words, $\pi_1(P) = \mathbb{Z}_2$.

Problems

20. Use van Kampen's theorem to calculate the fundamental group of the double torus by dividing the surface into two halves, each of which is a punctured torus. Do the calculation again, this time splitting the surface into a disc and the closure of the complement of the disc.

21. Show that the edge paths E_1, E_2 introduced in the proof of theorem (6.10) are equivalent.

22. Prove that the 'dunce hat' (Fig. 5.11) is simply connected using van Kampen's theorem.

23. Let X be a path-connected triangulable space. How does attaching a disc to X affect the fundamental group of X?

24. Let G be a finitely presented group. Construct a compact triangulable space which has fundamental group G.

6.5   Triangulating orbit spaces

Let K be a simplicial complex whose simplexes lie in $\mathbb{E}^n$. Then K is completely described once we know two things: the whereabouts of its vertices in $\mathbb{E}^n$ and which subsets of these vertices span simplexes. Let V denote the set of vertices of K, and S the collection of those subsets of V which span simplexes of K.

The pair $\{V,S\}$ is called the *vertex scheme* of K. The set V is finite and S has the following properties:

(a) Each element of V belongs to S. (A vertex is a 0-simplex.)
(b) If X belongs to S then any nonempty subset of X belongs to S. (Any face of a simplex of K is itself in K.)
(c) The sets in S are nonempty and have at most $m + 1$ elements for some non-negative integer m. (Take m to be the dimension of K.)

It is sometimes useful to be able to construct a simplicial complex by first specifying a finite nonempty set V, together with a collection S of subsets of V satisfying (a)–(c), and then 'realizing' the pair $\{V,S\}$ as the vertex scheme of a specific complex in some euclidean space. *Realization* means finding a simplicial complex K, and a bijection from V to the set of vertices of K, so that members of S correspond exactly to those sets of vertices which span simplexes. It should be clear that any two realizations of a given pair $\{V,S\}$ will be isomorphic complexes.

(6.14) Realization theorem. *Let* V *be a finite nonempty set and* S *a collection of subsets of* V *which satisfies properties* (a)–(c) *listed above. Then* {V,S} *can be realized as the vertex scheme of a simplicial complex.*

Proof. Suppose V has k elements, and let Δ be a $(k-1)$-simplex in $\mathbb{E}^{k-1}$. Then any one–one onto correspondence between the elements of V and the vertices of Δ realizes $\{V,S\}$ as the vertex scheme of a subcomplex of Δ. (In fact, no matter how large k is, we can always realize $\{V,S\}$ in $\mathbb{E}^{2m+1}$; see Problem 25.)

This method of constructing complexes will be used below to triangulate the orbit spaces of certain group actions, and again in Chapter 9 to define the dimension of a compact Hausdorff space.

It may happen that the space on which a group acts can be triangulated so that each group element induces a simplicial homeomorphism of the triangulation. In such a case, we shall say that the group action is *simplicial*. Fig. 6.17 gives a suitable triangulation for the antipodal action on S^2; take a regular octahedron inscribed inside the sphere and use radial projection π from the origin

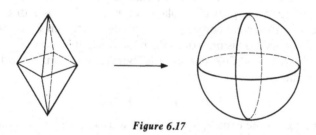

Figure 6.17

as the triangulating homeomorphism. The action is simplicial because the antipodal map $\phi:S^2 \to S^2$ induces a simplicial map $\pi^{-1}\phi\pi$ from the surface of the octahedron to itself. The three actions of $\mathbb{Z}_2$ on the torus described in Section 4.4, and the action of $\mathbb{Z}_p$ on S^3 which gives the Lens space $L(p,q)$ as orbit space, are other examples of simplicial actions. When we have a simplicial action, we shall show that the orbit space can be triangulated. Even better, we shall arrange things so that the natural projection is a simplicial map.

Suppose then we have a simplicial action of G on X, that is to say we assume the existence of a triangulation $h:|K| \to X$ such that $h^{-1}gh:|K| \to |K|$ is a simplicial homeomorphism for every element g of G. These homeomorphisms define an action of G on $|K|$, and to begin with we ignore X and work with this induced action on $|K|$.

We aim to triangulate the orbit space $|K|/G$. Using the projection $p:|K| \to |K|/G$, we define a pair $\{V,S\}$ as follows: the elements of V are the orbits (projections) of the vertices of K, and a subset $u_0,\ldots,u_k$ of V lies in S iff there exist vertices $v_0,\ldots,v_k$ of K which span a simplex of K and satisfy $p(v_i) = u_i$ for $0 \leqslant i \leqslant k$. The hypotheses of the realization theorem are easily

141

checked; realizing $\{V,S\}$ in some euclidean space produces a complex which we shall denote by K/G. Now p sends vertices of K to vertices of K/G, and if $v_0,\ldots,v_k$ span a simplex of K, then $p(v_0),\ldots,p(v_k)$ span a simplex of K/G. So p determines a simplicial map $s:|K| \to |K/G|$. Also, for any $x \in |K|$, $g \in G$ we have $sg(x) = s(x)$, so that s induces a function $\psi:|K|/G \to |K/G|$. The situation is best represented by means of a diagram

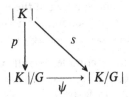

Clearly ψ is onto and, by theorem (4.1), is continuous iff s is continuous. But s is a simplicial map and therefore continuous. If ψ is one–one, it must be a homeomorphism by theorem (3.7), giving us a triangulation

$$\psi^{-1}:|K/G| \to |K|/G.$$

In general, ψ fails to be one–one. For example, in Fig. 6.17 the space $|K|/G$ is homeomorphic to the projective plane, whereas $|K/G|$ is a disc. However, if we replace K by its second barycentric subdivision K^2, then the corresponding map $\psi:|K|/G \to |K^2/G|$ is one–one (Problems 28, 29).

Now let $\hat{h}:|K|/G \to X/G$ denote the homeomorphism of orbit spaces induced by h, then

$$\hat{h}\psi^{-1}:|K^2/G| \to X/G$$

is a triangulation of the orbit space X/G. In addition, we have a commutative diagram

$$
\begin{array}{ccc}
|K^2| & \xrightarrow{\ h\ } & X \\
{\scriptstyle s}\downarrow & & \downarrow{\scriptstyle \pi} \\
|K^2/G| & \xrightarrow[\hat{h}\psi^{-1}]{} & X/G
\end{array}
$$

where π is the natural identification map. (To say the diagram commutes means simply that $\pi h(x) = \hat{h}\psi^{-1}s(x)$ for all $x \in |K^2|$.) The map s is simplicial, and it preserves the dimension of the simplexes of K^2, since two vertices of a simplex of K^2 cannot be mapped into one another by an element of G.

Problems

25. Suppose $\{V,S\}$ satisfies the hypotheses of the realization theorem, and label the elements of V as $v_1,\ldots,v_k$. If x_i denotes the point $(i,i^2,\ldots,i^{2m+1})$ of $\mathbb{E}^{2m+1}$, show that any $2m + 2$ of the points $x_1,\ldots,x_k$ are in general position, and deduce that the correspondence $v_i \leftrightarrow x_i$ can be used to realize $\{V, S\}$ in $\mathbb{E}^{2m+1}$.

26. By Problem 25 the vertex scheme of any one-dimensional complex can be realized in $\mathbb{E}^3$. Find a one-dimensional complex whose vertex scheme cannot be realized in $\mathbb{E}^2$.

27. Consider the antipodal action on S^2 and the triangulation shown in Fig. 6.17. Show that the map $\psi : |K|/G \to |K^1/G|$ is a homeomorphism, and draw the resulting triangulation of the projective plane.

28. Show that the map $\psi : |K|/G \to |K/G|$ is a homeomorphism iff the action of G on $|K|$ satisfies:

(a) The vertices of a 1-simplex of K never lie in the same orbit.

(b) If the sets of vertices $v_0, \ldots, v_k, a$ and $v_0, \ldots, v_k, b$ span simplexes of K, and if a, b lie in the same orbit, then there exists $g \in G$ such that $g(v_i) = v_i$ for $0 \leqslant i \leqslant k$ and $g(a) = b$.

29. Check that conditions (a) and (b) of Problem 28 are always satisfied if we replace K by its second barycentric subdivision.

6.6 Infinite complexes

So far, our simplicial complexes have contained only a finite number of simplexes. In order to deal with problems concerning noncompact spaces, we would like to relax this a little and allow certain infinite collections of simplexes to be complexes.

We shall insist that a complex be made up of simplexes which fit together nicely in some finite-dimensional euclidean space, and that the union of these simplexes form a closed subset of the euclidean space. Now if K is such a collection of simplexes in $\mathbb{E}^n$, then we can make a topological space $|K|$ out of their union by giving it the *induced* topology. An equally natural procedure is to take the simplexes of K separately, each with its topology induced from $\mathbb{E}^n$, and give their union the *identification* topology. We have seen in lemma (6.3) that these two procedures lead to the same topological space when K is finite. However, if we allow K to be infinite, then we may well obtain different answers. Indeed, a specific one-dimensional example where this happens is shown in Fig. 4.2.

Here is a tentative definition of an infinite complex, not the most general possible, but quite sufficient for our needs.

(6.15) Definition. *An infinite simplicial complex is an infinite collection of simplexes in some euclidean space $\mathbb{E}^n$ satisfying:*

(a) *if a simplex lies in the collection, then so does each of its faces;*

(b) *the simplexes in the collection fit together nicely;*

(c) *the union of all the simplexes is a closed subset of $\mathbb{E}^n$;*

(d) *the induced and identification topologies agree on the union of the simplexes.*

As a simple example of an infinite complex, take the strip $\{(x,y)\,|\,0 \leqslant y \leqslant 1\}$ in $\mathbb{E}^2$ divided up into triangles as shown in Fig. 6.18.

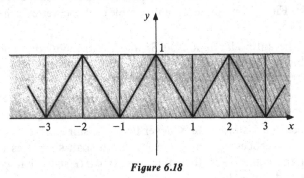

Figure 6.18

(6.16) Theorem. *Let* K *denote an infinite simplicial complex in* $\mathbb{E}^n$, *and let* $|K|$ *denote its polyhedron.*

(a) K *has finite dimension.*

(b) *The number of simplexes in* K *is countable.*

(c) K *is locally finite (that is to say, each vertex of* K *lies in only a finite number of simplexes).*

(d) *Each point of* $\mathbb{E}^n$ *has a neighbourhood which intersects at most a finite number of simplexes of* K .

Proof.

(a) Since K lies in $\mathbb{E}^n$, it cannot contain any simplexes of dimension greater than n, so the dimension of K is at most n.

(b) We prove that K contains only countably many simplexes by counting the set of barycentres of its simplexes. The topology on $|K|$ agrees with the identification topology, so this set of barycentres has no limit points in $\mathbb{E}^n$. There are therefore only finitely many barycentres inside any ball, centre the origin, of finite radius. Taking the union of the balls with integer radius shows the total number of barycentres to be countable.

(c) Suppose K is not locally finite and select a vertex v which is a vertex of infinitely many simplexes $A_1, A_2, \ldots$ of K. For each i, let x_i be a point which lies in the interior of A_i and whose distance from v is no more than 1. The set $\{x_i\}$ must have an accumulation point, say p, in $\mathbb{E}^n$, and since $|K|$ has the identification topology, p cannot lie in $|K|$. But this contradicts the fact that $|K|$ is supposed to be closed in $\mathbb{E}^n$.

(d) If $x \notin |K|$ then $\mathbb{E}^n - |K|$ is a neighbourhood of x ($|K|$ being closed) which does not meet $|K|$. If $x \in |K|$, select a vertex v of K for which $x \in \text{star}\,(v,K)$. Since $\text{star}\,(v,K)$ is open in $|K|$ we have $\text{star}\,(v,K) = O \cap |K|$ for some open subset O of $\mathbb{E}^n$, and O meets only a finite number of simplexes of K, by part (c).

We can define the notions of triangulation and simplicial map exactly as before. A space may now be noncompact and yet triangulable, though by

144

part (c) of theorem (6.16) it must be locally compact, in the sense that each of its points must have a compact neighbourhood. As we shall see, quite a few of the results of this chapter go through in this more general setting.

Note that our proof of the simplicial approximation theorem (6.7) made heavy use of the finiteness of the domain complex K (we needed $|K|$ to be compact), but the range L could have been infinite. Therefore we can show that the edge group of an infinite complex is isomorphic to the fundamental group of its polyhedron, and write down generators and relations for the group, exactly as in the finite case. However, the edge group need no longer be finitely presented.

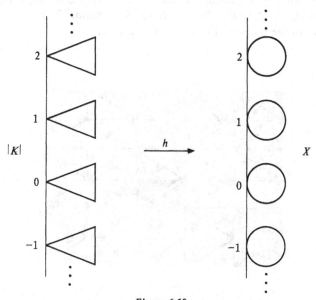

Figure 6.19

For example, if X consists of the real line with a circle attached at each integer, triangulated as in Fig. 6.19, its fundamental group is the free group on a countable number of generators. (We obtain a maximal tree containing all the vertices of K by taking the line together with two sides from each triangle. The remaining sides give the generators for $\pi_1(X)$, and there are no relations since K has no simplexes of dimension 2.) By shrinking the line to a point, we see that X has the homotopy type of a countable number of circles joined together at a single point.

If $\{V,S\}$ is the vertex scheme of an infinite complex, then V is a countable set, and in addition to properties (a)–(c) of Section 6.5 we have the property:

(d) Each element of V belongs to only a finite number of members of S. (K is locally finite.)

The realization theorem (6.14) remains true. Of course, the proof given earlier for the finite case cannot work, since K may now have an infinite number of vertices, but the method of Problem 25 goes through without difficulty.

We ask the reader to verify that we can allow infinite complexes in our work on triangulations of orbit spaces. This means we have many more examples of simplicial actions. The action of the integers on the real line by addition is simplicial; one has only to triangulate the real line as a one-dimensional complex by introducing a vertex at each integer. The action of a crystallographic group on the plane is simplicial: chop up a fundamental region into triangles and use the group action to tell you how to subdivide its translates. If the group is generated by a translation and a glide reflection acting at right angles, the fundamental region can be taken to be a rectangle and the resulting triangulation of the plane is illustrated in Fig. 6.20.

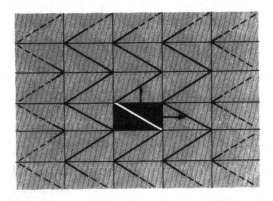

Figure 6.20

Let F be the free group on two generators x,y, set $V = F$, and agree that S consists of the elements of F together with pairs of elements g, h from F which have the property that $h^{-1}g$ is one of x, x^{-1}, y, y^{-1}. Let T denote the one-dimensional complex obtained by realizing $\{V,S\}$. Then T is connected because we can get from any element of F to any other by a sequence of operations, each of which amounts to multiplying on the right by one of x, x^{-1}, y, y^{-1}. Also, T must be simply connected, because any loop in T would lead to a non-trivial relation in F, and F is free. So T is a tree.

The complex T can in fact be realized in the plane, and we indicate how to do this in Fig. 6.21. Of course we have not been able to draw all of T! We shall call T the *universal television aerial*.

The action of F on itself by left multiplication (g sends h to gh) clearly induces a simplicial action of F on $|T|$, and the orbit space $|T|/F$ is the one-point union of two circles. Incidentally, this gives a second proof that the fundamental group of the one-point union of two circles is $\mathbb{Z} * \mathbb{Z}$, since the action of F on T satisfies the hypotheses of theorem (5.13). Now let H be a subgroup of F. Then H acts on T, and by our work in Section 6.5 we can triangulate the orbit space $|T|/H$ as a one-dimensional simplicial complex. The fundamental group of this orbit space is precisely H, by theorem (5.13), and we deduce that

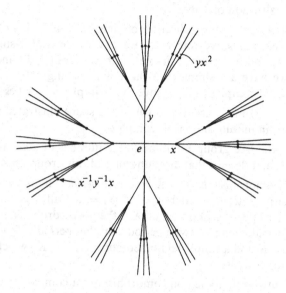

Figure 6.21

H is a free group since the fundamental group of a one-dimensional complex is always free. We have therefore proved the following result:

(6.17) Theorem. *Any subgroup of the free group on two generators is free.*

This is a special case of the Nielsen–Schreier theorem, which states that any subgroup of a free group is free. For more information see Problem 34.

We end this chapter with a generalization of theorem (5.13). Let G act simplicially on the path-connected triangulable space X, and let F be the normal subgroup of G generated by those elements which leave fixed at least one point of the space X.

(6.18) Theorem. *If* X *is simply connected, the fundamental group of the orbit space* X/G *is isomorphic to the factor group* G/F.

The proof is broken up into small steps in Problems 37–40. As an example, consider the crystallographic group generated by three half-turns illustrated in Fig. 4.5. For this action G and F coincide, since each of the three generators has a fixed point. Therefore the orbit space (the 2-sphere) is simply connected.

147

Problems

30. Find the triangulations of the sphere, torus, and Klein bottle which we obtain from the crystallographic groups shown in Fig. 4.5.

31. Check that the construction illustrated in Fig. 6.21 really can be carried out to produce a realization of T in $\mathbb{E}^2$.

32. Show that the following collection of simplexes in $\mathbb{E}^2$ is not a simplicial complex. For each positive integer n, we have a vertical 1-simplex joining $(1/n,0)$ to $(1/n,1)$ and a sloping 1-simplex with vertices $(1/n,0)$ and $(1/n + 1,1)$. In addition, we have a 1-simplex on the y axis joining $(0,0)$ to $(1,1)$. Do we obtain a simplicial complex if we remove the 1-simplex which lies on the y axis?

33. Can either the comb space (Fig. 5.10), or the space illustrated in Fig. 3.4, be triangulated by an infinite simplicial complex?

34. Show that the free group on a countable number of generators is a subgroup of $\mathbb{Z} * \mathbb{Z}$, and deduce that any subgroup of this group must be free.

35. Call a homeomorphism $h:X \to X$ *pointwise periodic* if for each point x of X there is a positive integer n_x such that $h^{n_x}(x) = x$. Call h *periodic* if $h^n = 1_X$ for some positive integer n. Show that if X is the polyhedron of a finite complex, and if h is simplicial, then pointwise periodic implies periodic. Find a connected infinite complex K and a simplicial homeomorphism of $|K|$ which is pointwise periodic but not periodic.

36. Does a pointwise periodic homeomorphism of a compact space have to be periodic? Be careful! (We comment that any pointwise periodic homeomorphism of a euclidean space is periodic, though this is hard to prove.)

37. Let G be a group of homeomorphisms of the space X. If N is a normal subgroup of G, show that G/N acts in a natural way on X/N and that X/G is homeomorphic to $X/N\big/G/N$. If F is the smallest normal subgroup of G which contains all the elements that have fixed points, show that G/F acts freely on X/F in the sense that only the identity element has any fixed points.

38. Suppose in addition to the conditions of Problem 37, X is a simply connected polyhedron, G acts simplicially, and X/G is triangulated so that the projection $p:X \to X/G$ is simplicial. Choose a vertex v of X as base point and define $\phi:G \to \pi_1(X/G,p(v))$ as follows. Given $g \in G$, join v to $g(v)$ by an edge path E in X; then $\phi(g)$ is the homotopy class of the edge loop $p(E)$. Show that ϕ is a homomorphism, and that each element of F goes to the identity under ϕ. Show also that ϕ is onto.

39. With the assumptions of Problem 38, show that X/F is simply connected, and that the action of G/F on X/F satisfies the hypotheses of theorem (5.13).

40. Now deduce theorem (6.18).

7. Surfaces

7.1 Classification

Results which allow one to classify completely a collection of objects are among the most important and aesthetically pleasing in mathematics. The fact that they are also rather rare adds even more to their appeal. As specific examples, we mention the classification of finitely generated abelian groups up to isomorphism in terms of their rank and torsion coefficients; that of quadratic forms in terms of the rank and signature of a form; and that of regular solids up to similarity by the number of edges of each face and the number of faces meeting at each vertex. It should be clear that we have no hope of classifying topological spaces up to homeomorphism, or even up to homotopy equivalence. We can, however, give a complete classification of closed surfaces.

A surface is *closed* if it is compact, connected, and has no boundary; in other words it is a compact, connected, Hausdorff space in which each point has a neighbourhood homeomorphic to the plane. When we say that we can classify such spaces, we mean that we can draw up a list (albeit infinite) of standard closed surfaces, all of which are topologically distinct, so that if we are presented with an *arbitrary* closed surface then it is homeomorphic to one on our list.

We recall (from Chapter 1) the statement of the classification theorem for closed surfaces:

(7.1) Classification theorem. *Any closed surface is homeomorphic either to the sphere, or to the sphere with a finite number of handles added, or to the sphere with a finite number of discs removed and replaced by Möbius strips. No two of these surfaces are homeomorphic.*

Adding a handle to the sphere means removing the interiors of a pair of disjoint discs, then attaching a cylinder by glueing its boundary circles to the edges of the two holes in the sphere, as illustrated in Fig. 7.1. When we add further handles, we do so on different parts of the sphere. Precisely where we add the handles (or Möbius strips) does not matter; we shall prove this carefully in Section 7.5. Notice how the handles are attached. If we mark arrows on the boundary circles of the cylinder, and on the edges of the holes in the sphere, to show how the identifications are to be made, and if the arrows on the cylinder go in the same direction, then those on the sphere will have opposite directions.

It is natural to ask what happens if, when we glue on a particular handle,

we reverse one of the arrows so that the two circles on the sphere are also given
the same sense. Take a disc in the sphere which contains, in its interior, the two
circles along which the cylinder is to be attached. Then the two possibilities are

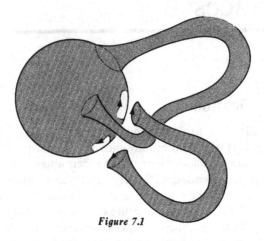

Figure 7.1

illustrated in Fig. 7.2. Now Fig. 7.2a is homeomorphic to the punctured torus,
and Fig. 7.2b to the punctured Klein bottle. Therefore *adding a handle* corresponds to removing a single open disc from the sphere and from the torus, and
glueing the two resulting boundary circles together. At first sight, we appear to
have a choice as to how we do this. For having marked an arrow on our circle
in the sphere, we can direct the boundary circle of the torus in two different
ways. However, there is a homeomorphism from the punctured torus to itself
which reverses the direction given to the boundary circle (Fig. 7.3a), and so
the two possibilities give homeomorphic answers.†

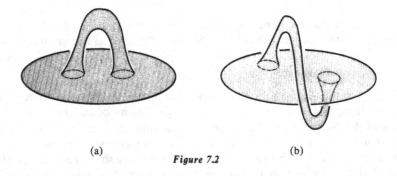

(a)   (b)

Figure 7.2

---

† The reader with a flair for precision will notice that we appeal to the following elementary
  proposition several times in this section. Given spaces X, $A \subseteq Y$, $B \subseteq Z$ together with maps
  $f:A \to X$, $g:B \to X$, and a homeomorphism $h:Z \to Y$ such that $h(B) = A$ and $fh = g$, then
  $X \cup_f Y$ and $X \cup_g Z$ are homeomorphic.

150

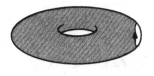

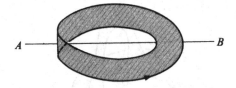

(a) Reflect in the plane of the paper (b) Rotate through π about axis AB

Figure 7.3

If the cylinder is attached the other way, as in Fig. 7.2b, then we must take a copy of the punctured Klein bottle and sew its boundary circle to the edge of a circular hole in the sphere. As above, the direction in which we glue the two circles is irrelevant. Now we know that the Klein bottle is the union of two Möbius strips along their boundary circles, or equivalently (Fig. 1.18) a cylinder with a Möbius strip glued to each of its boundary circles. So the punctured Klein bottle can be thought of as a disc with two holes punched in and a Möbius strip sewn into each hole. We have therefore shown that glueing a cylinder to the sphere 'in the wrong way' corresponds to removing *two* disjoint open discs from the sphere and sewing a Möbius strip into each of the resulting holes. Since the Möbius strip admits a homeomorphism which reverses the direction of its boundary circle (Fig. 7.3b), there is no ambiguity as to how we sew in these strips.

To complete our intuitive picture of how the sphere is modified when we add handles or sew in Möbius strips, we consider the possibility of doing a *mixture* of these operations. Suppose we have already replaced a disc by a Möbius strip, and we decide to add a handle. Then it does not matter how we do it; the operations shown in Figs 7.2a and 7.2b amount, in this situation, to the same thing. For call the Möbius strip M and the disc to which our cylinder is to be attached D. Run an arc from a point of the boundary circle of M to a point of the edge of D, and thicken it slightly to produce a band B, as in Fig. 7.4. Then $M \cup B \cup D$ is a Möbius strip and we are left to prove that the two spaces shown in Fig. 7.5 are homeomorphic. This is easily seen by cutting along the lines marked xy, when both become rectangles with a tube attached in precisely the same manner.

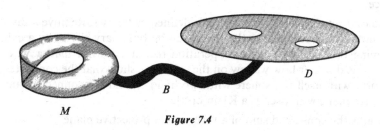

Figure 7.4

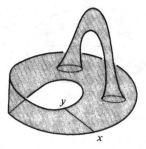
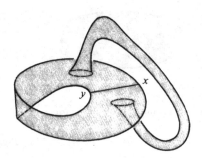

Figure 7.5

We summarize our discussion in the following result:

(7.2) Theorem. *Modifying the sphere by adding* m *handles and replacing* n (> 0) *disjoint discs by Möbius strips produces the same surface as replacing* 2m + n *disjoint discs by Möbius strips.*

Problems

1. The construction of a 'cross cap' is illustrated in Fig. 7.6. Show that punching a disc out of the sphere and adding a cross cap in its place gives a representation of the projective plane as a surface in $\mathbb{E}^3$ with self intersections.

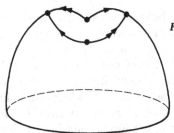
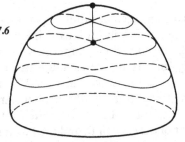

Figure 7.6

2. Let X consist of S^2 plus one extra point p. The neighbourhoods of the points of S^2 are the usual ones, and those of p are sets of the form $[U - \{(0,0,1)\}] \cup \{p\}$ where U is a neighbourhood of $(0,0,1)$ in S^2. Show that X is not Hausdorff, but is locally homeomorphic to the plane. Does it seem reasonable to call X a surface?

3. The *connected sum* of two surfaces is defined as follows. Remove a disc from each surface and connect up the two resulting boundary circles by a cylinder. Assuming this is a well-defined operation (i.e., it does not matter where we remove the discs, or how we sew on the cylinder), show that the connected sum of a torus with itself is a sphere with two handles, and the connected sum of a projective plane with itself is a Klein bottle.

4. What is the connected sum of a torus and a projective plane?

7.2 Triangulation and orientation

In order to make any headway at all we need to assume that our surfaces can be triangulated. That every compact surface admits a triangulation is a classical result of Rado proved in 1925. We shall not give a proof here;† we prefer to outline the idea but omit the details which are complicated and rather tedious.

Think of the problem of triangulating a closed surface S. Since S is compact and locally homeomorphic to the plane, we can find a finite number of closed discs in S whose union is all of S. To avoid annular regions between the discs, we agree to throw away any disc which lies entirely inside some other. Suppose (and this is the difficult step) we can arrange that the boundaries of these discs meet one another nicely, in the sense that if two meet then they do so in a finite number of points and arcs. *A priori* this need not be the case: think for example of the way a curve like $x \sin(1/x)$ meets the x axis near the origin. The union of the boundaries of our discs breaks up naturally as a set of arcs, and we introduce a vertex on the surface at each point where three or more arcs meet and at the midpoint of each arc (Fig. 7.7a). This produces most of the vertices and edges

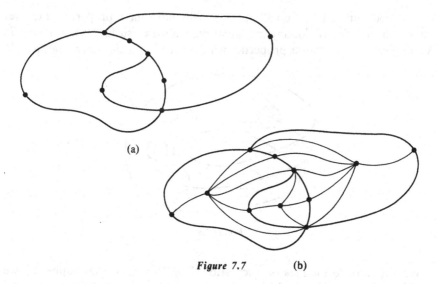

(a)

Figure 7.7   (b)

of our triangulation. To fill in the triangles, we note that S is nicely subdivided into patches which are homeomorphic to discs. All we have to do now is to triangulate each patch as a cone with apex some interior point, as in Fig. 7.7b. All this sounds temptingly easy, but we emphasize that finding a suitable covering by discs needs deep results, including a sharp version of the Jordan curve theorem.

---

† A short proof can be found in Doyle and Moran [25].

From now on we shall assume that all our surfaces can be triangulated. Let S be a closed surface and let $h:|K| \to S$ be a triangulation of S. As we might expect, K has some very nice properties. It has dimension 2; it is connected in the sense that any two of its vertices can be joined by an edge path; each of its edges is a face of exactly two triangles; and each vertex lies in at least three triangles which fit together to form a cone with apex the given vertex and base a polygonal simple closed curve (as in Fig. 7.8). A triangulation constructed by

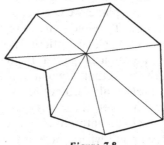

Figure 7.8

the method outlined above will automatically have these properties, but they can be verified directly for *any* triangulation of a closed surface (see Problem 7). A complex with these four properties will be called a *combinatorial surface*.

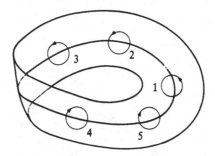

Figure 7.9

We now turn to the idea of orientation. Fig. 7.9 shows what happens if we translate a small circle, to which we have given a sense of rotation, once round the central circle of a Möbius strip. The effect on the circle is to reverse its sense. For this reason, surfaces which contain a Möbius strip are said to be *nonorientable*. A surface like the torus which does not contain a Möbius strip, and for which the operation of translating a small, oriented circle round a simple closed curve always preserves the sense of the circle, is called *orientable*.

We can approach this idea in a different way by making use of the fact that our surfaces are triangulable. There are two ways to orient, or give a sense of rotation to, a triangle. The two possibilities are shown in Fig. 7.10 and can be specified either by drawing arrows on the triangle or, if we want to be completely precise, by appropriately ordering the vertices of the triangle. Of course, if we choose the ordering v_0, v_1, v_2 to specify a particular orientation, then we must

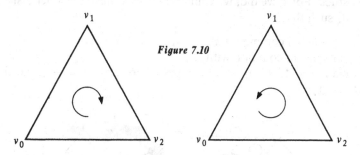

Figure 7.10

agree that the cyclic variations v_2, v_0, v_1 and v_1, v_2, v_0 represent the *same* orientation.

This idea works for a simplex of any dimension. Let A be a general simplex and consider two orderings of its vertices to be equivalent if they differ by an even permutation. There are precisely two equivalence classes (unless A consists of a single vertex), each of which is called an *orientation* of A. Of course, a vertex can only be oriented in one way. Suppose now that we have chosen an orientation for A by ordering its vertices in some way, say as $v_0, v_1,..., v_k$, and let B be the face of A determined by deleting v_i. Then the vertices of B are automatically ordered. If i is even, the orientation of B specified by this ordering is called the orientation *induced* from A. If i is odd, we take the other orientation of B as that induced from A. The simplest case is the sense of direction given to each edge of a triangle by a choice of orientation of the triangle. It is easy to see that this definition depends only on the orientation of A, and not on the particular ordering of the v_i chosen to represent this orientation.

We say that a combinatorial surface K is *orientable* if it is possible to orient all the triangles of K in a compatible manner. That is to say, in such a way that two adjacent triangles always induce *opposite* orientations on their common edge. Fig. 7.11 illustrates this definition. We leave the reader to experiment

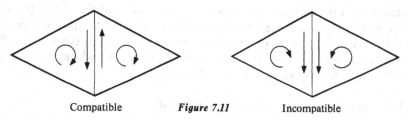

Compatible   *Figure 7.11*   Incompatible

with triangulations of the torus and Klein bottle: for the torus there is never any difficulty in orienting the triangles compatibly; in the case of the Klein bottle one always gets stuck.

155

If $h:|K| \to S$ is a triangulation of an orientable surface S, then the complex K must be orientable. For choose a 2-simplex in K and orient it; induce compatible orientations to its neighbours and continue round the complex. We never get stuck. For if we did, we could find a sequence of distinct 2-simplexes $A_1, A_2, \ldots, A_k$ such that:

(a) A_i has an edge in common with A_{i+1}, $1 \le i \le k-1$;
(b) A_k has an edge in common with A_1;
(c) the orientations of A_i, A_{i+1} are compatible for $1 \le i \le k-1$, but those of A_k, A_1 are not.

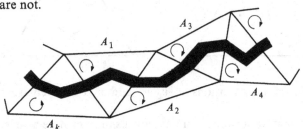

Figure 7.12

Join the barycentre of A_i to the midpoints of the edges $A_{i-1} \cap A_i$ and $A_i \cap A_{i+1}$ by straight lines, where A_{i-1} is interpreted as A_k when $i = 1$, and A_{i+1} as A_1 when $i = k$. This gives a simple closed polygonal path in $|K|$ which we thicken slightly to obtain a strip (Fig. 7.12). Since the orientations of the triangles are compatible, with the exception of those of A_1 and A_k, the strip is a Möbius strip in $|K|$. This contradicts the assumption that S is an orientable surface.

We shall return to the question of orienting surfaces in the next chapter and show that if S is a closed surface which can be triangulated by an orientable combinatorial surface, then *any* other triangulation of S must also be orientable.

We have used the idea of thickening a polygonal curve in a combinatorial surface. Since this type of process will be needed quite frequently, we end this section with a couple of lemmas designed to make it quite precise. Let K be a combinatorial surface, and let L be a one-dimensional subcomplex of K. To *thicken* L we first barycentrically subdivide K twice, and then form the subcomplex consisting of those simplexes of K^2 which meet L, together with all their faces (Fig. 7.13). Finally we take the polyhedron of this subcomplex. We obtain in this way a closed neighbourhood of $|L|$ in $|K|$ which, it is not too hard to prove, has the same homotopy type as $|L|$.

(7.3) Lemma. *Thickening a tree always gives a disc.*

Proof. Proceed by induction on the number of vertices in the tree T. If T consists of a single vertex v, then thickening T gives precisely the union of those simplexes of K^2 which have v as a vertex: this is called the *closed star* of v in K^2 and written $\overline{star}(v, K^2)$. It should be clear that K^2 is a combinatorial surface, and that $\overline{\text{star}}\,(v,K^2)$ is a disc. If T has n vertices, choose an 'end' vertex v, that is, one which belongs to only one† edge E of T. Remove this edge to produce a tree T_1 with one less vertex. Thickening T_1 gives a disc D by assumption, and in

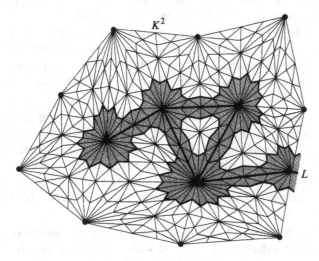

Figure 7.13

order to thicken T, all we have to do is to add to this disc the two closed stars $A = \overline{\text{star}}\,(\hat{E},K^2)$, $B = \overline{\text{star}}\,(v,K^2)$. Now A and B are both discs; moreover, A and D meet along their boundaries in an arc, as do A and B. Two applications of lemma (2.11) show that $D \cup A \cup B$ is a disc.

(7.4) Lemma. *Thickening a simple closed polygonal curve gives either a cylinder or a Möbius strip.*

Proof. Remove an edge E from the curve to give a tree in K. Thickening this tree gives a disc D, by lemma (7.3). To thicken the original curve, we need to take the union of D with the closed star of the barycentre $\hat{E}$ in K^2. Now $\overline{\text{star}}$ $(\hat{E}, K^2)$ and D meet along their boundaries in two disjoint arcs. If we glue $\overline{\text{star}}\,(\hat{E}, K^2)$ to D along one of these arcs we obtain a disc, by lemma (2.11). It remains only to identify two disjoint arcs in the boundary of this new disc. Define a homeomorphism from the disc to a rectangle so that the arcs go to a pair of opposite sides. (The homeomorphism is defined on the two arcs first, then extended over the rest of the boundary, and finally extended over the whole disc by means of lemma (2.10).) We now have to identify a pair of opposite sides of a rectangle: there are only two ways of doing this, and they give the cylinder and the Möbius strip.

---

† Such a vertex exists, since if each vertex lies in two edges it is easy to find a loop in T. But T is a tree and therefore cannot contain a loop.

Problems

5. Suppose we want to triangulate a surface which has a boundary. How does the definition of a combinatorial surface need to be adjusted?

6. Let K be a combinatorial surface. Show that the triangles of K can be labelled $T_1,\ldots,T_s$ in such a way that T_i always has an edge in common with at least one of $T_1,\ldots,T_{i-1}$. Now build a model for the surface $|K|$ in the form of a regular polygon in the plane, which has an even number of sides, and whose sides have to be identified *in pairs* in some way.

7. If $h:|K| \to S$ is a triangulation of a closed surface, show that K must be a combinatorial surface. This requires a little patience. First use a connectivity argument locally to show that K cannot have dimension 1. Then prove K cannot contain a simplex of dimension greater than 2, and that every edge of K lies in precisely two triangles, using methods like those of theorem (5.23). Finally, check that the triangles of K which contain a particular vertex fit together as in Fig. 7.8.

8. Let G be a finite group which acts as a group of homeomorphisms of a closed surface S in such a way that the only element with any fixed points is the identity. Show that the orbit space S/G is a closed surface. Show that S may be orientable, yet S/G nonorientable. If S/G is orientable, does S have to be so? A group action for which only the identity element has any fixed points is said to be *fixed-point free*, and the group in question is said to act *freely*.

9. Let K be an orientable combinatorial surface, orient all its triangles in a compatible manner, and let $h:|K| \to |K|$ be a simplicial homeomorphism. Suppose there is a triangle A, oriented by the ordering u, v, w of its vertices, whose image $h(A)$ occurs with the orientation $h(u)$, $h(v)$, $h(w)$ induced by h. Prove that the same must hold for any other triangle of K, and call h *orientation-preserving*. Give an example of an orientable combinatorial surface and a simplicial homeomorphism which is not orientation-preserving.

10. Let K be an orientable combinatorial surface. If G acts simplicially on $|K|$, if the action is fixed-point free, and if each element of G is orientation-preserving, show that the complex K^2/G is an orientable combinatorial surface.

7.3 Euler characteristics

Let S be a closed surface. We know from our remarks on triangulation that we may replace S, up to homeomorphism, by the polyhedron of a combinatorial surface K. For the remainder of this section we shall work with the space $|K|$.

If L is a finite simplicial complex of dimension n, we define its *Euler characteristic* to be

$$\chi(L) = \sum_{i=0}^{n} (-1)^i \alpha_i$$

where α_i is the number of i-simplexes in L. So if L happens to be a combinatorial surface, $\chi(L)$ is the number of vertices minus the number of edges plus the number of triangles, and the Euler characteristic of a graph† is the number of vertices minus the number of edges. As mentioned in Chapter 1, $\chi(L)$ is a topological invariant of the space $|L|$. The proof of this fact will be given in Chapter 9; we do not need it here.

(7.5) Lemma. $\chi(\Gamma) \leqslant 1$ *for any graph* Γ, *and equality occurs if and only if* Γ *is a tree.*

Proof. If Γ is a tree, it is easy to show $\chi(\Gamma) = 1$ by induction on the number of vertices. If Γ is not a tree, then it must contain a loop. Removing an edge from the loop leaves Γ connected and increases the Euler characteristic, since the number of edges decreases by 1 while the number of vertices remains constant. By repetition of this process we eventually convert Γ into a tree. Therefore $\chi(\Gamma) < 1$.

Suppose now that K is a combinatorial surface, and that T is a maximal tree in K. We know from lemma (6.11) that T contains all the vertices of K. Construct a graph Γ, called the *dual* to T, by realizing the following vertex scheme: the vertices of Γ are the barycentres of triangles of K, and two such barycentres span a 1-simplex of Γ if and only if the corresponding triangles meet in an edge of K which does not lie in T. (We refer the reader back to Fig. 1.5.)

If we take the first barycentric subdivision Γ^1, then we can think of it as the subcomplex of K^1 consisting of all those simplexes which do not meet T. We make use of this representation of Γ^1 to show that Γ *is connected*. Thicken T and do the same for Γ (in other words, form the union of those simplexes of K^2 which meet Γ). Call the resulting spaces $N(T)$ and $N(\Gamma)$ respectively. We know from lemma (7.3) that $N(T)$ is a disc, and it is not hard to check the following:

(a) $N(T) \cup N(\Gamma) = |K|$;
(b) $N(T)$ and $N(\Gamma)$ intersect in precisely the boundary circle of $N(T)$;
(c) Γ is a connected complex if and only if $N(\Gamma)$ is a connected space.

Now any two points x, y of $N(\Gamma)$ can be joined by a path in $|K|$. Let p, q be the first and last points where this path intersects the boundary of $N(T)$. Follow the given path from x to p, go round the boundary circle of $N(T)$ from p to q, then take the given path again from q to y. This joins x to y by a path in $N(\Gamma)$. Therefore Γ is a graph, by (c) above.

Note that Γ need not be a tree. Fig. 7.14 shows a triangulation of the torus, and a choice of tree T, for which Γ has the homotopy type of a bouquet of two circles.

---

† A connected one-dimensional complex.

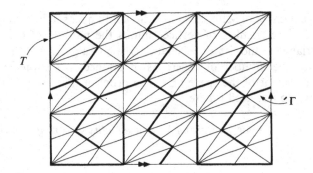

Figure 7.14

(7.6) Lemma. $\chi(K) \leqslant 2$ *for any combinatorial surface* K.

Proof. Choose a maximal tree T in K and construct its dual graph Γ as above. Observe that $\chi(K) = \chi(T) + \chi(\Gamma)$, since all the vertices of K lie in T, Γ has an edge for every edge of K not in T, and the number of vertices in Γ is precisely the number of triangles in K. Therefore $\chi(K) \leqslant 2$ by lemma (7.5).

(7.7) Theorem. *The following are equivalent for any combinatorial surface* K:
(a) *Every simple closed polygonal curve in* $|K|$ *which is made up of edges of* K^1 *separates* $|K|$;
(b) $\chi(K) = 2$;
(c) $|K|$ *is homeomorphic to the sphere.*

Proof. Suppose (a) is satisfied. Choose a maximal tree T in K and let Γ be its dual. We claim that Γ is also a tree giving $\chi(K) = \chi(T) + \chi(\Gamma) = 2$. For if not, Γ must contain a loop, and by assertion this loop separates $|K|$. But each component of the complement of this loop must contain a vertex of T, contradicting the fact that T is connected and disjoint from Γ. Therefore Γ is indeed a tree.

If $\chi(K) = 2$, then $\chi(\Gamma)$ must be 1 and so Γ is a tree. Therefore $|K|$ is the union of two discs $N(T)$ and $N(\Gamma)$ along their boundary circles, giving the sphere.

Finally, the proof of theorem (5.20) tells us that any simple closed curve on the sphere separates the sphere. This completes the chain of implications (a) $\Rightarrow$ (b) $\Rightarrow$ (c) $\Rightarrow$ (a).

We shall need two further results concerning the Euler characteristic; we relegate their proofs to the exercises which follow.

(7.8) Lemma. *Let* K, L *be simplicial complexes which intersect in a common subcomplex, then* $\chi(K \cup L) = \chi(K) + \chi(L) - \chi(K \cap L)$.

(7.9) Lemma. *The Euler characteristic is left unchanged by barycentric subdivision.*

160

Problems

11. Prove lemma (7.8).

12. Prove lemma (7.9) by induction on the number of simplexes in the complex.

13. Deduce from Problem 12 that the Euler characteristic of a graph Γ is a topological invariant of $|\Gamma|$.

14. Let K be a finite complex. If G acts simplicially on $|K|$, and if the action is fixed-point free, show that

$$\chi(K) = |G| \cdot \chi(K^2/G)$$

where $|G|$ denotes the number of elements in G.

15. Let K be a combinatorial surface. Make a model for $|K|$ in the plane, as in Problem 6, and let J denote the boundary curve of the resulting regular polygon. Identifying the edges of J in pairs according to the prescription for building $|K|$, gives a graph Γ in K. Show that $\chi(K) = \chi(\Gamma) + 1$, then deduce lemma (7.6) from lemma (7.5).

16. Continuing from Problem 15, if Γ has an edge, one end of which is not joined to any of the other edges, show there must be two edges in J which have a vertex in common and which are 'folded together' about this common vertex when we form Γ from J. Hence give a second proof that $\chi(K) = 2$ implies $|K| \cong S^2$.

7.4 Surgery

We now begin our attack on the classification theorem by showing how to modify a given combinatorial surface in such a way as to *increase* its Euler characteristic. The modification involves cutting out part of the surface and replacing it by something else, and is quite aptly called 'surgery'. We have just seen that a combinatorial surface has Euler characteristic less than or equal to 2, and that equality occurs precisely when the underlying space is homeomorphic to the sphere. Consequently, we shall be able to convert every surface into the sphere by a finite number of our so-called surgeries.

Fig. 7.15 illustrates the type of modification we have in mind for the double

Figure 7.15

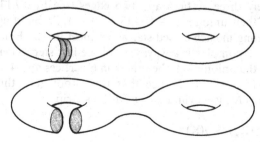

torus. We begin with a simple closed curve which does not separate the surface into two pieces, and thicken it to obtain a cylinder. Doing surgery along the curve involves removing the interior of this cylinder and filling in each of the two resulting holes with a disc. The result is a surface homeomorphic to the torus. A further surgery will give us the sphere. Of course, if we begin with a nonorientable surface, then thickening the curve may well give a Möbius strip. In this case, we remove the interior of the strip to give a compact surface with boundary a single circle, then close up the surface again by capping off this circle with a disc.

We have drawn our picture without reference to any particular triangulation for the sake of clarity, and because this is how we visualize a surgery. However, we emphasize that we do need to work throughout with *combinatorial* surfaces in order to have the Euler characteristic available as a tool.

Let K be a combinatorial surface in $\mathbb{E}^n$, and let L be a simple closed polygonal curve which is a subcomplex of K and which does not separate $|K|$. Form the second barycentric subdivision K^2 and thicken L, calling the resulting *complex* N. By lemma (7.4), we know that $|N|$ is either a cylinder or a Möbius strip. Let M be the subcomplex of K^2 which is complementary to N, that is to say M consists of those simplexes of K^2 which do not meet L, together with all their faces. One possibility is that thickening L gives a cylinder: then $|M|$ is a compact surface with boundary consisting of two circles, and we label the subcomplexes which triangulate these circles by L_1, L_2. We now form the new combinatorial surface

$$K_* = M \cup CL_1 \cup CL_2$$

where the apexes of the cones CL_1, CL_2 are points of $\mathbb{E}^{n+1} - \mathbb{E}^n$ which lie on opposite sides of $\mathbb{E}^n$. The other possibility is that thickening L gives a Möbius strip. In this case, $|M|$ has a single circle as boundary; we call the subcomplex triangulating this circle L_1 and define K_* to be $M \cup CL_1$. In both cases we say that K_* is obtained from K by doing *surgery along* L.

(7.10) Lemma. $\chi(N) = 0$

Proof. Examine carefully the proof of lemma (7.4). N is built up of the closed stars, $\overline{\text{star}}\,(v, K^2)$ where $v \in L^1$. Now the closed star of a vertex in a combinatorial surface clearly has Euler characteristic 1. If two of these closed stars meet, they do so in precisely three vertices and two edges (see Fig. 7.13), so the Euler characteristic of their union is $1 + 1 - (3 - 2) = 1$. So build up N by walking round L and adding in each closed star as we meet it. The Euler characteristic of the resulting subcomplex is always 1 until the last step, when we add a star which intersects the union of all the others in 6 vertices and 4 edges. Therefore $\chi(N) = 1 + 1 - (6 - 4) = 0$. Notice that this proof goes through independently of whether $|N|$ is a cylinder or a Möbius strip.

(7.11) Theorem. $\chi(K_*) > \chi(K)$

Proof. If thickening L gives a cylinder,

$$\chi(K_*) = \chi(M) + \chi(CL_1) + \chi(CL_2) - \chi(L_1) - \chi(L_2)$$
$$= \chi(M) + 2$$

If thickening L gives a Möbius strip,

$$\chi(K_*) = \chi(M) + \chi(CL_1) - \chi(L_1)$$
$$= \chi(M) + 1$$

Also, combining previous lemmas,

$$\chi(K) = \chi(K^2) = \chi(M) + \chi(N) - \chi(M \cap N) = \chi(M)$$

This completes the argument.

If K is a combinatorial surface, and if $|K|$ is homeomorphic to the sphere, then we shall call K a *combinatorial sphere*.

(7.12) Corollary. *Any combinatorial surface can be changed into a combinatorial sphere by a finite number of surgeries.*

Proof. If $\chi(K) = 2$, then K is a combinatorial sphere and we have nothing to do. If $\chi(K) < 2$, there is a simple closed polygonal curve in K^1 which does not separate $|K|$, by theorem (7.7). So replace K by K^1 and do surgery along such a curve. The result is a new combinatorial surface whose Euler characteristic is larger than that of K. Continuing in this way, we eventually produce a combinatorial surface with Euler characteristic 2.

We shall need a slight refinement of the above argument. Each time we do a surgery, we create either one or two discs on the surface, and we would like to ensure that the curves along which we do our subsequent surgeries avoid these discs. If we are unlucky and find ourselves with a curve, along which we wish to do surgery, and which runs through a disc, then we simply shrink the disc into the interior of one of its triangles, and hence off the curve. Of course we must be careful when shrinking our disc, not to move any of the other discs onto the curve. The following lemma allows us to realize this shrinking process.

(7.13) Lemma. *Let K be a combinatorial surface, D a disc which is a subcomplex of K, and A a triangle of D. There is a homeomorphism $h:|K| \to |K|$ which satisfies $h(D) = \overline{star}(\hat{A}, K^2)$ and which is the identity on all simplexes of K that do not meet D.*

The *idea* of the proof is very simple. Thicken the boundary of D to produce a slightly larger disc D_1, then shrink D onto $\overline{star}(\hat{A}, K^2)$ inside D_1, keeping all of $|K| - D_1$ fixed. Further details can be found in Problems 19–21. If L is a polygonal curve in K which intersects D and along which we need to do surgery, then before carrying out the surgery we replace D by $\overline{\text{star}}\,(\hat{A}, K^2)$ and replace K by K^2.

We can now prove one half of the classification theorem:

(7.14) Theorem. *Every closed surface is homeomorphic to one of the standard ones.*

Proof. Given a closed surface S, triangulate it and do surgery on the resulting combinatorial surface until it becomes a combinatorial sphere. The triangulation is no longer needed and we can forget it. After the surgeries, we are left with a sphere which has a number of disjoint discs marked on it. To recapture S, all we have to do is reverse each of the surgeries; this involves either removing a pair of discs and attaching a cylinder in their place, or replacing a single disc by a Möbius strip.

If our original surface S is orientable, it does not contain any Möbius strips, therefore reversing a surgery must always amount to removing a pair of discs and adding a *handle* in their place. So we obtain a sphere with handles. If S is nonorientable, then both types of operation can occur. But we know from Section 7.1 that, in this case, removing two discs and sewing on a cylinder is always equivalent to replacing each of the discs by a Möbius strip. Therefore we obtain a sphere with a finite number of Möbius strips sewn in.

Problems

17. The straight lines shown in Fig. 7.16 represent three simple closed curves in the Klein bottle. Thicken each curve, decide whether the result is a cylinder or a Möbius strip, and describe the effect of doing surgery along the curve.

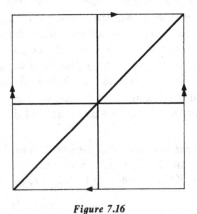

Figure 7.16

18. Show that the surface illustrated in Fig. 7.17 is homeomorphic to one of the standard ones using the procedure of theorem (7.14).

164

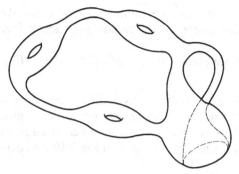

Figure 7.17

19. Let $X \supset Y \supset Z$ be three concentric discs in the plane. Find a homeomorphism from X to itself which is the identity on the boundary circle of X and which throws Y onto Z.

20. Suppose we have two discs in the plane, both of which are bounded by polygonal curves, and one of which lies in the interior of the other. Prove that the region between them is homeomorphic to an annulus. (The best hint we can give is Fig. 7.18, plus a reminder that any polygonal simple closed curve in the plane bounds a disc.)

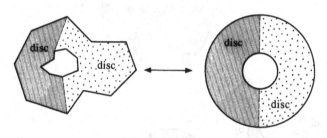

Figure 7.18

21. With the notation of lemma (7.13) and Problem 19, find a homeomorphism $h: D_1 \rightarrow X$ such that $h(D) = Y$ and $h(\overline{\text{star}}\,(\hat{A}, K^2)) = Z$. Now prove lemma (7.13).

7.5 Surface symbols

Write $H(p)$ for the sphere with p handles added, and $M(q)$ for the sphere with q Möbius strips sewn in. Two questions remain unanswered.

Question 1. Are the surfaces $H(p)$ and $M(q)$ well defined? In other words, if we take a sphere and add some handles (or Möbius strips), and if we take a second copy of the sphere and add the same number of handles (or Möbius strips) but to different parts of the sphere, are the resulting surfaces homeomorphic?

Question 2. Are the standard surfaces S^2, $H(1)$, $M(1)$, $H(2)$, $M(2)$, $H(3)$,... topologically distinct?

We shall deal with these two questions by constructing models for our standard surfaces. Consider the orientable case first. Suppose we are given a sphere with two handles attached. For each of the handles we choose a pair of simple closed curves which wind round it once, as shown in Fig. 7.19. The curves are all based

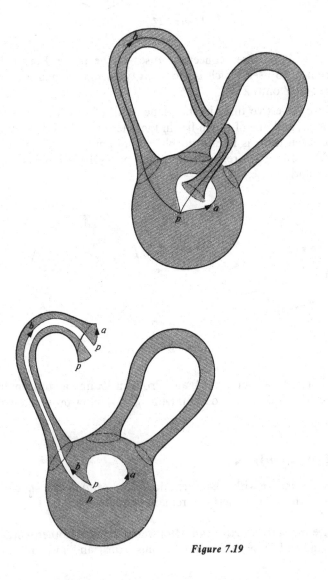

Figure 7.19

at the same point and are otherwise disjoint. Suppose we now cut the surface along the curves labelled a, b in the directions indicated by the arrows. Then we can open the surface out to become a rectangle with a handle attached in its interior. A further pair of cuts, along the curves c and d, produces a model in the form of an eight-sided polygon with its sides labelled appropriately (Fig. 7.20).

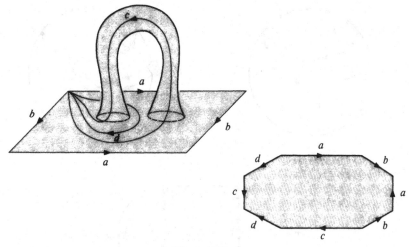

Figure 7.20

Now the original surface is completely defined by the way in which we identify the edges of this polygon in pairs, and this information is efficiently stored in the so-called *surface symbol* obtained by reading round the polygon clockwise and listing the labels on its edges as they occur, adding a superscript -1 to each edge whose arrow points anticlockwise. So the surface symbol of the sphere with two handles attached is $aba^{-1}b^{-1}cdc^{-1}d^{-1}$.

By increasing the number of cuts, we can clearly produce a model for the sphere with p handles attached in the form of a $4p$-sided polygon whose edges are to be identified in pairs according to the symbol $a_1b_1a_1^{-1}b_1^{-1}a_2b_2a_2^{-1}b_2^{-1}$ $\ldots a_pb_pa_p^{-1}b_p^{-1}$. Since two surfaces which have the same surface symbol are quite clearly homeomorphic, we have disposed of Question 1 for orientable surfaces.

In the nonorientable case, we cut each Möbius strip along a curve which cuts across it once, as in Fig. 7.21. We leave the reader to check that if we have q Möbius strips, then we obtain a $2q$-sided polygon with surface symbol $a_1a_1a_2a_2\ldots a_qa_q$. Again, we have answered Question 1 in the affirmative.

To show that the surfaces S^2, $H(1)$, $M(1)$, $H(2)$, $\ldots$ are all topologically distinct, we shall calculate their fundamental groups. To illustrate the method, which uses van Kampen's theorem (6.13), we again choose to work with $H(2)$. Removing an open disc from $H(2)$ produces a space which deformation retracts onto the one-point union of four circles; the fundamental group of this space

167

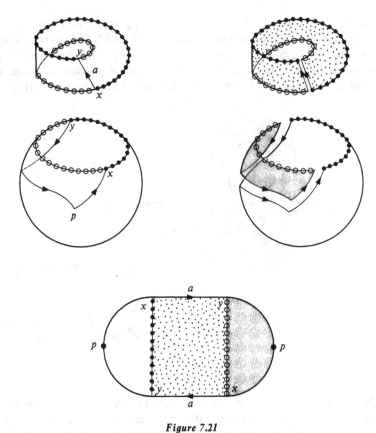

Figure 7.21

is $\mathbb{Z} * \mathbb{Z} * \mathbb{Z} * \mathbb{Z}$ with generators a, b, c, d represented by our original four loops. Now if we use C to denote the boundary circle of this space, then the loop $aba^{-1}b^{-1}cdc^{-1}d^{-1}$ is quite clearly homotopic to a generator of $\pi_1(C)$. So van Kampen's theorem gives

$$\pi_1(H_2) \cong \{a,b,c,d \mid aba^{-1}b^{-1}cdc^{-1}d^{-1} = e\}$$

The same sort of argument shows

$$\pi_1(H(p)) \cong \{a_1,b_1,\ldots,a_p,b_p \mid \prod_{i=1}^{p} a_i \, b_i \, a_i^{-1} \, b_i^{-1} = e\}$$

and

$$\pi_1(M(q)) \cong \{a_1,a_2,\ldots,a_q \mid \prod_{i=1}^{q} a_i^2 = e\}$$

We also know, of course, that S^2 is simply connected.

If we now abelianize each of these groups, in other words, form the quotient of each group by its commutator subgroup, then $\pi_1(H(p))$ becomes the free

168

abelian group $\mathbb{Z} \times \mathbb{Z} \times \ldots \times \mathbb{Z}$ on $2p$ generators, and $\pi_1(M(q))$ the abelian group generated by q elements $x_1, x_2 \ldots, x_q$ subject to $(x_1 x_2 \ldots x_q)^2 = e$. Changing to the new set of generators $x_1 x_2 \ldots x_q, x_2, x_3, \ldots, x_q$, we see that this latter group is $\mathbb{Z}_2 \times \mathbb{Z} \times \ldots \times \mathbb{Z}$, there being $q - 1$ infinite cyclic factors. Since no two of the abelianized groups are isomorphic, we conclude that no two of our standard surfaces are homeomorphic. This completes our classification of closed surfaces.

$H(p)$ is called the standard orientable surface of *genus* p, and $M(q)$ the standard nonorientable surface of genus q. *A closed surface is completely determined once we know its genus and whether or not it is orientable.*

At this point we recommend the reader to work through the alternative (historically much earlier) proof of the classification theorem for closed surfaces given in Massey [9].

Much current work in topology centres on the study of manifolds, the higher-dimensional analogues of surfaces. A *manifold of dimension n* (*n*-manifold for short) is a second-countable Hausdorff space, each point of which has a neighbourhood homeomorphic to $\mathbb{E}^n$. The spaces $\mathbb{E}^n$, S^n, P^n are all n-manifolds; $S^3 \times S^1$ is a closed 4-manifold ('closed' meaning it is compact and connected); any open subset of an n-manifold is itself an n-manifold, so $GL(n)$ is a manifold of dimension n^2; $SO(n)$ is a closed manifold of dimension $\frac{1}{2}n(n-1)$; finally, the Lens spaces $L(p,q)$ are all examples of closed 3-manifolds. Despite a great deal of progress, many basic questions remain unanswered. The most important is the famous *Poincaré conjecture*. When posed as a question, it asks if every closed, simply connected manifold of dimension 3 is homeomorphic to S^3.

Problems

22. Are the surfaces shown in Fig. 7.22 homeomorphic?

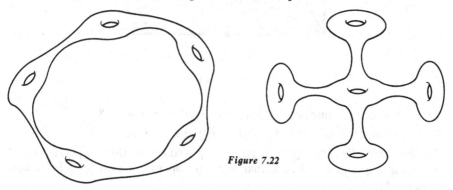

Figure 7.22

23. What happens if we remove the interiors of two disjoint discs from a closed surface, then glue the two resulting boundary circles together?

24. Use the classification theorem to show that the operation of connected sum (Problem 3) is well defined.

25. Assuming every compact surface can be triangulated, show that the boundary of a compact surface, if nonempty, consists of a finite number of disjoint circles.

26. Show that any compact connected surface is homeomorphic to a closed surface from which the interiors of a finite number of disjoint discs have been removed.

27. What is the fundamental group of the space obtained by punching k holes in the sphere?

28. Write $H(p,r)$ for $H(p)$ with the interiors of r disjoint discs removed, and $M(q,s)$ for $M(q)$ with s discs similarly removed. Show that $H(p,r)$ can be obtained from a $(4p + 3r)$-sided polygonal region in the plane by glueing up its edges according to the surface symbol

$$a_1 b_1 a_1^{-1} b_1^{-1} \ldots a_p b_p a_p^{-1} b_p^{-1} x_1 y_1 x_1^{-1} \ldots x_r y_r x_r^{-1}.$$

(Figure 7.23 is supplied as a hint.)

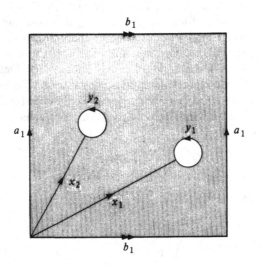

Figure 7.23

29. Find a surface symbol for $M(q,s)$, as defined in Problem 28.

30. Calculate the fundamental groups of $H(p,r)$ and $M(q,s)$.

31. Show that $H(p,r) \cong H(p',r')$ implies $p = p'$ and $r = r'$; that $M(q,s) \cong M(q',s')$ implies $q = q'$ and $s = s'$; and that there are no values of p, q, r, s for which $H(p,r) \cong M(q,s)$.

32. Define the *genus* of a compact connected surface to be the genus of the closed surface obtained on capping off each boundary circle with a disc. Show that a compact connected surface is completely determined by whether or not it is orientable, together with its genus and its number of boundary circles.

33. Identify the surfaces shown in Fig. 7.24. Can you suggest a general result from these two pictures?

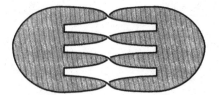 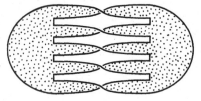

Figure 7.24

8. Simplicial Homology

8.1 Cycles and boundaries

If we wish to distinguish between the sphere and the torus, we have already seen one way of doing so using the fundamental group. Any loop in the sphere can be continuously shrunk to a point, in other words the sphere is simply connected, whereas this is not the case for the torus. The fundamental group is a very valuable tool, but it has a significant defect. Remember that the fundamental group of a polyhedron depends only on the 2-skeleton of the underlying complex, making it ideal for studying questions which are essentially two-dimensional (say distinguishing between two surfaces), but leaving it impotent in the face of a problem such as showing that S^3 and S^4 are not homeomorphic.

In an attempt to overcome this difficulty, we shall associate to each finite simplicial complex K a collection of groups $H_q(K), q = 0,1,2,\ldots$, called the simplicial homology groups of K. These groups will be defined using the simplicial structure of K, but they will turn out to depend only on the homotopy type of the polyhedron $|K|$, allowing us to define the homology groups of any compact triangulable space. Each $H_q(K)$ is a finitely generated abelian group, and is to be thought of as in some sense measuring $(q + 1)$-dimensional holes in the space $|K|$. For example, the group $H_4(S^4)$ will be shown to be non-trivial, verifying our feeling (when we look at S^4 in $\mathbb{E}^5$) that S^4 has a five-dimensional hole.

The construction of the homology groups of a complex is quite complicated, and for this reason we attempt to provide a little motivation in what follows. We can distinguish between the sphere and the torus in a rather different manner from that suggested above. Every simple closed curve drawn on the sphere separates it, and therefore forms the boundary of a region on the sphere. This is not so for the torus: Fig. 8.1 shows three simple closed curves on the torus

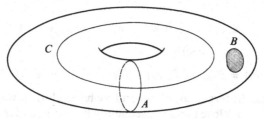

Figure 8.1

only one of which, the curve B, bounds a piece of the surface. In order to recognize the fact that the torus has holes in it, we would like some way of working with simple closed curves that ignores those which bound part of the surface.

It is important to realize that a curve may bound part of a surface and yet not be null homotopic. For example, the boundary circle of the punctured torus bounds the whole surface, but we know that it represents a nontrivial element in the fundamental group.

For reasons which will become apparent later, we choose to work with *oriented* polygonal curves in some fixed triangulation K of the torus, denoting orientation as usual by arrows on the edges of the curves. If an edge has vertices v, w, then the symbol (v,w) will denote this edge oriented in the direction from v to w. In a similar manner, if u, v, w are the vertices of a triangle of K, then (u,v,w) denotes this triangle oriented by the ordering u, v, w of its vertices; so $(u,v,w) = (v,w,u) = (w,u,v)$. We denote a change of orientation by a minus sign, thus $(w,v) = -(v,w)$ and $(v,u,w) = -(u,v,w)$. The *boundary* of the oriented edge (v,w) is defined to be

$$\partial(v,w) = w - v$$

and the boundary of the oriented triangle (u,v,w) is

$$\partial(u,v,w) = (v,w) + (w,u) + (u,v)$$

Note that the boundary of (u,v,w) is the sum of its edges, each taken with the orientation induced by the given orientation of the whole triangle.

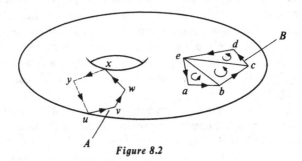

Figure 8.2

If we now think of an oriented curve such as A in Fig. 8.2 as the sum of its oriented edges

$$A = (u,v) + (v,w) + (w,x) + (x,y) + (y,u)$$

and define its boundary linearly by

$$\partial A = \partial(u,v) + \partial(v,w) + \partial(w,x) + \partial(x,y) + \partial(y,u)$$

then of course all the terms cancel out and we have a formal way of recognizing that a curve like A is closed and consequently has no boundary. Now consider the oriented curve B. It is also closed and, in addition, it encloses three of the triangles of K. If we orient these triangles as indicated, write their union as

$$(e,a,b) + (e,b,c) + (e,c,d)$$

and compute the boundary of this, we obtain

$$\partial(e,a,b) + \partial(e,b,c) + \partial(e,c,d)$$

$$= (a,b) + (b,e) + (e,a) + (b,c) + (c,e) + (e,b) + (c,d) + (d,e) + (e,c)$$

$$= (a,b) + (b,e) + (e,a) + (b,c) + (c,e) - (b,e) + (c,d) + (d,e) - (c,e)$$

$$= (a,b) + (b,c) + (c,d) + (d,e) + (e,a)$$

$$= B$$

making precise the fact that B bounds a piece of the torus.

We now consider arbitrary linear combinations $\lambda_1(u_1,v_1) + \ldots + \lambda_k(u_k,v_k)$ of oriented edges of K, with integer coefficients,† which have no boundary in the sense that $\lambda_1 \partial(u_1,v_1) + \ldots + \lambda_k \partial(u_k,v_k)$ vanishes. Such an expression will be called a *one-dimensional cycle* of K. We have lost some geometric content in doing this (after all, 'five times a simplex' does not mean very much!), but we do have the advantage that our 1-cycles form an abelian group under the addition

$$\Sigma\lambda_i(u_i,v_i) + \Sigma\mu_i(u_i,v_i) = \Sigma(\lambda_i + \mu_i)(u_i,v_i)$$

We denote this group by $Z_1(K)$.

An oriented, simple closed polygonal curve in K, when thought of as the sum of its oriented edges, is a particularly simple sort of 1-cycle and may be referred to as an elementary 1-cycle. It is an easy exercise (which we recommend to the reader) to verify that $Z_1(K)$ is generated by these elementary cycles.

Thinking back to the curve B above, we say that a one-dimensional cycle is a *bounding cycle* if we can find a linear combination of oriented triangles whose boundary is the given cycle. The bounding cycles rather obviously form a subgroup $B_1(K)$ of $Z_1(K)$, and we wish to ignore these. Consequently, we form the quotient group

$$H_1(K) = Z_1(K)/B_1(K)$$

which we call the *first homology group* of K.

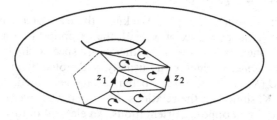

Figure 8.3

---

† Keeping in mind that $\lambda(u,v)$ always means the same as $(-\lambda)(v,u)$.

175

Two cycles whose difference is a bounding cycle represent the same element of $H_1(K)$ and are said to be homologous. For example, the two cycles z_1, z_2 shown in Fig. 8.3 are homologous, since $z_1 - z_2$ is the boundary of the tube (with its triangles oriented as shown) between them.

As we shall see later, $H_1(K)$ turns out to be isomorphic to $\mathbb{Z} \oplus \mathbb{Z}$, and we can represent the generators by elementary cycles z_1, z_2, where z_1 winds once round the torus longitudinally, and z_2 winds once round meridianally. So any other one-dimensional cycle is homologous to a linear combination of these two. For example, in the triangulation shown in Fig. 8.4, the 'diagonal' cycle z is homologous to $z_1 + z_2$ since $z_1 + z_2 - z$ bounds half of the torus.

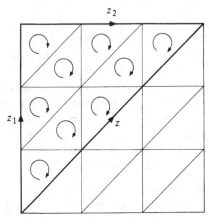

Figure 8.4

In order to motivate the definition of the first homology group, we have worked with a specific triangulation of the torus. However, it is clear that our construction makes sense for any simplicial complex K. Even better, it generalizes easily to provide a homology group $H_q(K)$ for each nonnegative integer q, as we shall see in what follows.

8.2 Homology groups

Let K be a finite simplicial complex. We know there are precisely two different ways of orienting each simplex of K, with the exception that a vertex can be oriented in only one way. A simplex together with a specific choice of orientation will be called an *oriented simplex* and usually be denoted by the symbol σ or τ.

We define $C_q(K)$ to be the free abelian group generated by the oriented q-simplexes of K, subject to the relations $\sigma + \tau = 0$ whenever σ and τ are just the same simplex with opposite orientations. An element of this group is called a *q-dimensional chain*, and $C_q(K)$ itself will be referred to as the qth chain group of K. Note that $C_q(K)$ is free abelian with rank equal to the number of q-simplexes in K.

A q-chain can be thought of as a linear combination $\lambda_1\sigma_1 + \ldots + \lambda_s\sigma_s$ of oriented q-simplexes of K with integer coefficients, provided we remember that $\lambda(-\sigma)$ and $(-\lambda)\sigma$ always mean the same thing, where $-\sigma$ as usual stands for σ with its orientation reversed.

We shall often want to define homomorphisms on these chain groups, and in doing so our approach will always be the same: specify the value of the homomorphism on each generator of $C_q(K)$, that is to say on an arbitrary oriented q-simplex σ of K; check that the relations $\sigma + (-\sigma) = 0$ are preserved; then extend linearly to the other elements.

A good example is the *boundary homomorphism*. The boundary of an oriented q-simplex is defined to be the $(q-1)$-chain determined by the sum of its $(q-1)$-dimensional faces, each taken with the orientation induced from that on the whole simplex.

We need a little more notation in order to produce a formula from which we can compute boundaries. If a q-simplex has vertices $v_0, \ldots, v_q$, the symbol $(v_0, \ldots, v_q)$ means this simplex oriented by the given ordering. Therefore

$$(v_0, \ldots, v_q) = \text{sign}\, \theta\, (v_{\theta(0)}, \ldots, v_{\theta(q)})$$

for any permutation θ of $0, \ldots, q$, where $\text{sign}\, \theta = +1\,(-1)$ if θ is an even (odd) permutation. According to the above description the boundary of the oriented q-simplex $(v_0, \ldots, v_q)$ is

$$\partial(v_0, \ldots, v_q) = \sum_{i=0}^{q} (-1)^i (v_0, \ldots, \hat{v}_i, \ldots, v_q),$$

where $(v_0, \ldots, \hat{v}_i, \ldots, v_q)$ is shorthand for the oriented $(q-1)$-simplex obtained by deleting the vertex v_i. One easily checks that changing the orientation of σ changes the induced orientation on each of its faces, so $\partial\sigma + \partial(-\sigma)$ is zero. Therefore ∂ determines a homomorphism

$$\partial : C_q(K) \to C_{q-1}(K)$$

In the special case when $q = 0$, we define the boundary of a single vertex to be zero and set $C_{-1}(K) = 0$.

Thinking back to our work in Section 8.1, it is natural for us to call the kernel of $\partial : C_q(K) \to C_{q-1}(K)$ the group of q-cycles of K, and denote it by $Z_q(K)$.

8.1 Lemma. *The composition $C_{q+1}(K) \xrightarrow{\partial} C_q(K) \xrightarrow{\partial} C_{q-1}(K)$ is the zero homomorphism.*

Proof. We need only check that $\partial^2 = \partial \circ \partial$ gives zero when applied to any

177

oriented $(q + 1)$-simplex of K. Now

$$\partial^2(v_0,\ldots,v_{q+1}) = \partial \sum_{i=0}^{q+1} (-1)^i (v_0,\ldots,\hat{v}_i,\ldots,v_{q+1})$$

$$= \sum_{i=0}^{q+1} (-1)^i \sum_{j=i+1}^{q+1} (-1)^{j-1} (v_0,\ldots,\hat{v}_i,\ldots,\hat{v}_j,\ldots,v_{q+1})$$

$$+ \sum_{i=0}^{q+1} (-1)^i \sum_{j=0}^{i-1} (-1)^j (v_0,\ldots,\hat{v}_j,\ldots,\hat{v}_i,\ldots,v_{q+1})$$

All the terms in this expression cancel in pairs, since each oriented $(q - 1)$-simplex $(v_0,\ldots,\hat{v}_i,\ldots,\hat{v}_j,\ldots,v_{q+1})$ appears twice, the first time with sign $(-1)^{i+j-1}$ and the second time with the opposite sign $(-1)^{i+j}$.

If we write $B_q(K)$ for the image of $\partial : C_{q+1}(K) \to C_q(K)$, the above lemma shows us that $B_q(K)$ is a subgroup of $Z_q(K)$. We call $B_q(K)$ the group of *bounding q-cycles*.

The *qth homology group of K* is now defined to be

$$H_q(K) = Z_q(K)/B_q(K)$$

The element of $H_q(K)$ determined by a q-cycle z will be called the *homology class* of z and written $[z]$. Two q-cycles whose difference is a bounding q-cycle have the same homology class and will be called *homologous* cycles.

A homology group $H_q(K)$ is by its very definition a finitely generated abelian group. Therefore it can be written in the form $F \oplus T$, where F is a finitely generated free abelian group (in other words, the direct sum of a finite number of copies of $\mathbb{Z}$), and T is a finite abelian group. The elements of T are precisely those elements of the homology group which have finite order, and they are called *torsion* elements. The rank of F, that is, the number of summands when we express F as a sum of cyclic groups, is called the *qth Betti number†* of K and denoted by β_q.

Problems

1. Check that changing the orientation of a simplex changes the induced orientation on each of its faces.

2. Show the elementary 1-cycles, mentioned in Section 8.1, generate $Z_1(K)$ for any complex K.

3. Take the triangulation of the Möbius strip shown in Fig. 6.2, orient one of the triangles, then go round the strip orienting each triangle in a manner compatible with the one preceding it. (Of course, when you get back to where you started the orientations do not match up.) What is the boundary of the two-dimensional chain formed by taking the sum of these oriented triangles?

---

† After the Italian mathematician Enrico Betti (1823–92).

4. Let K be the complex shown in Fig. 6.4, assuming the identifications are made so that $|K|$ is a torus. Orient the triangles of K in such a way that if two have an edge in common, their orientations are not compatible. Now take the sum of all the oriented triangles and compute its boundary.

5. As for Problem 4, but this time orient all the triangles compatibly, with the exception of one of them which is given the 'wrong' orientation.

6. Triangulate the 'dunce hat' (Fig. 5.11) in some way and decide whether or not there are any 2-cycles.

7. Show that any cycle of K is a bounding cycle of the cone on K.

8. Triangulate S^n so that the antipodal map is simplicial and induces a triangulation of P^n. If n is odd, find an n-cycle in this triangulation of P^n. What difficulties arise when n is even?

9. Triangulate the Möbius strip in a simple way so that its centre circle is a subcomplex. Orient the boundary circle of the strip, and the centre circle, calling the resulting elementary 1-cycles z_1, z respectively. Show that z_1 is homologous to either $2z$ or $-2z$.

10. Suppose $|K|$ is homeomorphic to the torus with the interiors of three disjoint discs removed. Orient each boundary circle of $|K|$ and let z_1, z_2, z_3 be the resulting elementary 1-cycles of K. Show that $[z_3] = \lambda[z_1] + \mu[z_2]$ where $\lambda = \pm 1$, $\mu = \pm 1$. Do we have the same result if we replace the torus by the Klein bottle?

8.3 Examples

In this section we shall calculate one or two homology groups. The methods used will be rather primitive, and purposely so because any systematic calculation of these groups would take us too far afield. Our aim is to reach the stage where we can present some significant applications of homology theory as quickly as possible, and with a minimum of fuss. For a more sophisticated approach, see Maunder [18].

Example 1. Let K be the complex shown in Fig. 8.5.

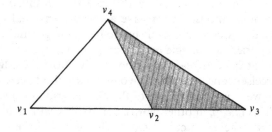

Figure 8.5

179

The vertices v_1, v_2, v_3, v_4 generate $Z_0(K) = C_0(K)$, and $C_1(K)$ can be thought of as the free abelian group generated by the oriented 1-simplexes (v_1, v_2), (v_1, v_4), (v_2, v_3), (v_2, v_4), (v_3, v_4). So $B_0(K) = \partial C_1(K)$ is generated by $v_2 - v_1$, $v_4 - v_1$, $v_3 - v_2$, $v_4 - v_2$, $v_4 - v_3$, and we see that v_1, v_2, v_3, v_4 all determine the same homology class. Therefore $H_0(K) = Z_0(K)/B_0(K)$ is an infinite cyclic group generated by $[v_1]$.

The group $Z_1(K)$ is generated by the elementary 1-cycles, and by inspection there are six such, namely

$$z_1 = (v_1, v_2) + (v_2, v_4) + (v_4, v_1)$$

$$z_2 = (v_2, v_3) + (v_3, v_4) + (v_4, v_2)$$

$$z_3 = (v_1, v_2) + (v_2, v_3) + (v_3, v_4) + (v_4, v_1)$$

plus $-z_1$, $-z_2$, and $-z_3$. Since $z_3 = z_1 + z_2$, we see that $Z_1(K) \cong \mathbb{Z} \oplus \mathbb{Z}$ with generators z_1, z_2. Our complex K has only one two-dimensional simplex, so $C_2(K)$ is an infinite cyclic group generated by (v_2, v_3, v_4). This means that $B_1(K) = \partial C_2(K)$ is generated by $\partial(v_2, v_3, v_4) = z_2$. So the first homology group $H_1(K)$ is isomorphic to $\mathbb{Z}$ and generated by $[z_1]$.

Finally, there are no 2-cycles, and no simplexes of dimension greater than 2, therefore $H_q(K) = 0$ for $q \geqslant 2$.

Example 2. If two vertices v, w of a complex K lie in the same component of $|K|$, then they are homologous. For we can join v to w by an edge path $v \, v_1 \, v_2 \ldots v_k \, w$, in which no two consecutive vertices are equal, and then check that $w - v$ is the boundary of the 1-chain $(v, v_1) + (v_1, v_2) + \ldots + (v_k, w)$. We leave the reader to convince himself that vertices which lie in different components of $|K|$ are not homologous, and that an integer multiple of a single vertex can never be a boundary, thus proving the following result:

(8.2) Theorem. $H_0(K)$ *is a free abelian group whose rank is the number of components of* $|K|$.

Example 3. Let K be a triangulation of the torus. If we orient all the 2-simplexes of K compatibly, take their sum, and compute the boundary of this sum, then each edge of the triangulation occurs exactly twice in the result, once with each of its two possible orientations. So we have a two-dimensional cycle. It is elementary to check that any other 2-cycle has to be an integer multiple of this one. (For suppose the oriented triangle (a, b, c) occurs in a 2-cycle with coefficient λ, then $\lambda(b, c)$ automatically appears in its boundary. Now the edge spanned by b and c lies in precisely one other triangle of K whose third vertex we denote by d. The only way we can rid ourselves of the above term $\lambda(b, c)$ is to orient this adjacent triangle as (d, c, b), in other words, compatibly with (a, b, c), and include it in our cycle with the same coefficient λ. Going round the complex in this way, we see we must orient all the simplexes compatibly, and give them all the same coefficient.) Since there are no 3-simplexes in a triangulation of the torus, there are no bounding cycles, and therefore $H_2(K)$ is isomorphic to $\mathbb{Z}$.

If we now change to a triangulation of the punctured torus, there are no 2-cycles, since even if we include all the triangles oriented compatibly as above, when we compute the boundary we are left with those edges which form the boundary of the hole in the torus. So the second homology group is zero.

The second homology group of a triangulation of the Klein bottle is also zero. Again, there are no 2-cycles, but this time for a different reason. The Klein bottle being nonorientable, there is no way to orient compatibly all the 2-simplexes of a triangulation.

Notice how the second homology group very nicely distinguishes between the torus, which is orientable, and the Klein bottle, which is not.

Example 4. Suppose we have a complex K which is a cone, in other words K is isomorphic to a complex of the form CL where the dimension of L is one less than that of K. Let v be the unique vertex of K which does not lie in L, usually called the apex of K.

A cone is always connected, so $H_0(K) \cong \mathbb{Z}$, by theorem (8.2). Now assume $q > 0$ and define a homomorphism $d: C_q(K) \to C_{q+1}(K)$ as follows. If $\sigma = (v_0, \ldots, v_q)$ is an oriented q-simplex of K which happens to lie in L, define $d(\sigma) = (v, v_0, \ldots, v_q)$; otherwise set $d(\sigma) = 0$. Clearly $d(\sigma)$ depends only on the orientation of σ (and not on the particular ordering of its vertices chosen to represent this orientation), and $d(\sigma) + d(-\sigma) = 0$ in $C_{q+1}(K)$. So d gives a homomorphism from $C_q(K)$ to $C_{q+1}(K)$. Now check that

$$\partial d(\sigma) = \sigma - d\partial(\sigma)$$

for any oriented q-simplex σ. (For example, if σ lies in L then

$$\partial d(\sigma) = \partial(v, v_0, \ldots, v_q)$$

$$= (v_0, \ldots, v_q) + \sum_{i=0}^{q} (-1)^{i+1} (v, v_0, \ldots, \hat{v}_i, \ldots, v_q)$$

$$= \sigma - d\partial(\sigma)$$

The other case is left to the reader.) So if z is a q-cycle of K, we have $\partial d(z) = z - d\partial(z) = z$. This shows every q-cycle to be a bounding cycle, and therefore $H_q(K) = 0$ for $q > 0$.

Example 5. Let Δ^{n+1} denote an $(n+1)$-simplex, $n > 0$, together with all its faces, thought of as a simplicial complex, and let Σ^n denote those simplexes which lie in the boundary of Δ^{n+1}. So $|\Sigma^n|$ is homeomorphic to S^n. Now Σ^n and Δ^{n+1} have precisely the same simplexes up to and including dimension n. Also, the definition of the qth homology group does not involve simplexes of dimension greater than $q + 1$, and therefore $H_q(\Sigma^n) \cong H_q(\Delta^{n+1})$ for $0 \leqslant q \leqslant n - 1$. But Δ^{n+1} is a cone, so by Example 4 we have $H_0(\Sigma^n) \cong \mathbb{Z}$ and $H_q(\Sigma^n) = 0$

181

for $1 \leqslant q \leqslant n - 1$. (Remember we have assumed $n > 0$. Σ^0 consists of two points and so $H_0(\Sigma^0) \cong \mathbb{Z} \oplus \mathbb{Z}$ by Example 2.)

Since Σ^n has no $(n + 1)$-simplexes, $H_n(\Sigma^n) = Z_n(\Sigma^n) = Z_n(\Delta^{n+1})$. And, since $H_n(\Delta^{n+1}) = 0$, we have $Z_n(\Delta^{n+1}) = B_n(\Delta^{n+1}) = \partial C_{n+1}(\Delta^{n+1})$. The latter group is clearly infinite cyclic, therefore $H_n(\Sigma^n) \cong \mathbb{Z}$. We can obtain a generator by orienting all the n-simplexes of Σ^n compatibly. Of course, $H_q(\Sigma^n) = 0$ for $q > n$.

Once we have verified the topological invariance of homology groups, we will be able to refer to the groups $H_q(\Sigma^n)$ as the homology groups of the n-sphere.

Example 6. Edge loops and elementary 1-cycles look remarkably similar, so we are not surprised to find a close connection between the edge group of a complex and its first homology group.

Suppose $|K|$ is connected and choose a vertex v to act as base point. Any edge loop $\alpha = vv_1v_2...v_kv$ gives us a 1-cycle $z(\alpha) = (v,v_1) + (v_1,v_2) + ... + (v_k,v)$ if we agree to omit (v_i,v_{i+1}) whenever $v_i = v_{i+1}$. If two edge loops differ by a single operation of the type used to define equivalence of edge loops, they clearly determine homologous cycles. So the correspondence $\alpha \mapsto z(\alpha)$ gives us a function $\phi : E(K,v) \rightarrow H_1(K)$. It follows from the definition of ϕ that it is a homomorphism. We shall show that ϕ is onto and that the kernel of ϕ is the commutator subgroup of $E(K,v)$. Remembering that the edge group $E(K,v)$ is isomorphic to the fundamental group of $|K|$, we will have the following result:

(8.3) Theorem. *If $|K|$ is connected, abelianizing its fundamental group gives the first homology group of* K.

To show ϕ is onto, we need only prove that the homology class of each elementary 1-cycle lies in the image of ϕ. Now an elementary 1-cycle is just an oriented simple edge loop thought of as the sum of its oriented edges, say $z_1 = (w_1,w_2) + (w_2,w_3) + ... + (w_s,w_1)$. If we join v to w_1 by an edge path γ and set $\alpha = \gamma w_1 w_2...w_s \gamma^{-1}$, then $z(\alpha) = z_1$ as required.

Since $H_1(K)$ is an abelian group, the kernel of ϕ must contain the commutator subgroup of $E(K,v)$. To complete our proof, we must show that if α is an edge loop for which $z(\alpha)$ is a bounding cycle, then $\{\alpha\}$ lies in the commutator subgroup of $E(K,v)$. As above, write $\alpha = vv_1v_2...v_k v$, and suppose

$$z(\alpha) = \partial(\lambda_1 \sigma_1 + ... + \lambda_l \sigma_l)$$

where the σ_i are oriented 2-simplexes of K. Suppose $\sigma_i = (a_i, b_i, c_i)$ and for each i choose an edge path γ_i joining v to a_i. The edge loop $\gamma_i a_i b_i c_i \gamma_i^{-1}$ is equivalent to the trivial edge loop v, and therefore so is the product

$$\beta = \prod_{i=1}^{l} (\gamma_i a_i b_i c_i \gamma_i^{-1})^{\lambda_i}$$

giving $\{\alpha \beta^{-1}\} = \{\alpha\}$. Note that

$$z(\gamma_i a_i b_i c_i \gamma_i^{-1}) = \partial(a_i, b_i, c_i)$$

and hence $z(\alpha\beta^{-1}) = 0$. Now the only way an edge loop can map to the zero 1-cycle under $\alpha \mapsto z(\alpha)$ is if, whenever an oriented edge (a,b) occurs n times in it, then (b,a) also occurs n times. Recall the homomorphism $\theta : E(K,v) \to G(K,L)$ defined in theorem (6.12). Under θ the equivalence class of a loop such as $\alpha\beta^{-1}$ will map to a product of group elements in which each group element occurs the same number of times as its inverse. Therefore if we first apply θ and then abelianize $G(K,L)$, our element $\{\alpha\beta^{-1}\}$ will map to zero. But θ is an isomorphism, and so $\{\alpha\beta^{-1}\} = \{\alpha\}$ must lie in the commutator subgroup of $E(K,v)$, completing the argument.

Suppose now that K is a combinatorial surface. Then $H_0(K) \cong \mathbb{Z}$ by theorem (8.2), and in order to find $H_1(K)$ all we have to do is abelianize the fundamental group of $|K|$. This was done at the end of Chapter 7, and we remind the reader of the result:

$$H_1(K) \cong \begin{cases} 0 & \text{if } |K| \text{ is the sphere} \\ 2g\mathbb{Z} & \text{if } |K| \text{ is an orientable surface of genus } g \\ (g-1)\mathbb{Z} \oplus \mathbb{Z}_2 & \text{if } |K| \text{ is a nonorientable surface of} \\ & \text{genus } g \end{cases}$$

Also the arguments given in Example 3 above show that $H_2(K)$ is $\mathbb{Z}$ if the combinatorial surface K is orientable, and 0 if not. Accepting for the moment that homology groups are topological invariants, we can rewrite this as

$$H_2(K) = \begin{cases} \mathbb{Z} & \text{if } |K| \text{ is an orientable surface} \\ 0 & \text{if not.} \end{cases}$$

Problems

11. Calculate the homology groups of the following complexes: (a) three copies of the boundary of a triangle all joined together at a vertex; (b) two hollow tetrahedra glued together along an edge; (c) a complex whose polyhedron is homeomorphic to the Möbius strip; (d) a complex which triangulates the cylinder.

12. What are the homology groups of a tree?

13. Show that any graph has the homotopy type of a bouquet of circles, and suggest a formula for the first Betti number of the graph.

14. Calculate the homology groups of a triangulation of the 'dunce hat'.

15. Finish the computation $\partial d(\sigma) = \sigma - d\partial(\sigma)$ of Example 4.

16. Calculate the homology groups of a triangulation of the sphere with k holes.

17. If $|K|$ is homeomorphic to the standard orientable surface $H(p,r)$ with r holes, show that the first Betti number of K is given by

$$\beta_1 = 2p + r - 1.$$

What is the second Betti number of K?

18. What are the Betti numbers of K if $|K|$ is homeomorphic to $M(q,s)$ (defined in Problem 28 of Chapter 7)?

19. What is the nth Betti number of a triangulation of P^n?

8.4   Simplicial maps

Let K, L be complexes and $s:|K| \to |L|$ a *simplicial* map. Using s we shall construct a homomorphism $s_q : C_q(K) \to C_q(L)$ for each q.

Remember that a simplicial map takes simplexes linearly onto simplexes, but that it may decrease the dimension of a simplex. Given an oriented q-simplex $\sigma = (v_0,\ldots,v_q)$ of K, we define $s_q(\sigma)$ to be the oriented q-simplex $(s(v_0),\ldots,s(v_q))$ of L if all the vertices $s(v_0),\ldots,s(v_q)$ are distinct, and we set $s_q(\sigma) = 0$ otherwise. This determines a homomorphism from $C_q(K)$ to $C_q(L)$, since clearly $s_q(-\sigma) = -s_q(\sigma)$.

We claim that s_q, in turn, induces a homomorphism $s_{q*} : H_q(K) \to H_q(L)$. In order to prove this, we must show that s_q takes cycles of K to cycles of L, and bounding cycles to bounding cycles. This is most efficiently done using the following lemma:

(8.4) Lemma. $\partial s_q = s_{q-1}\partial : C_q(K) \to C_{q-1}(L)$, *that is to say, the following diagram commutes*:

$$
\begin{array}{ccc}
C_q(K) & \xrightarrow{\ s_q\ } & C_q(L) \\
\big\downarrow{\scriptstyle\partial} & & \big\downarrow{\scriptstyle\partial} \\
C_{q-1}(K) & \xrightarrow{\ s_{q-1}\ } & C_{q-1}(L)
\end{array}
$$

Proof. We show that $\partial s_q(\sigma) = s_{q-1}\partial(\sigma)$ for any oriented q-simplex $\sigma = (v_0,\ldots,v_q)$ of K. This is clear if all the vertices $s(v_0),\ldots,s(v_q)$ are distinct. If not, suppose $s(v_j) = s(v_k)$, where $j < k$. By definition we have $s_q(\sigma) = 0$, so $\partial s_q(\sigma) = 0$. Now

$$s_{q-1}\partial(\sigma) = \sum_{i=0}^{q} (-1)^i s_{q-1}(v_0,\ldots,\hat{v}_i,\ldots v_q).$$

Examining the terms in this sum, if i is not j or k, then

$$s_{q-1}(v_0,\ldots,\hat{v}_i,\ldots,v_q) = 0.$$

The two remaining terms are

$$(-1)^j s_{q-1}(v_0,\ldots,\hat{v}_j,\ldots,v_q) \text{ and } (-1)^k s_{q-1}(v_0,\ldots,\hat{v}_k,\ldots,v_q).$$

These are nonzero only if v_j and v_k are the only vertices of σ identified by s, and in this case the two terms cancel because

$$s_{q-1}(v_0,\ldots,\hat{v}_j,\ldots,v_q) = (s(v_0),\ldots,\widehat{s(v_j)},\ldots,s(v_q))$$
$$= (-1)^{k-j-1}(s(v_0),\ldots,\widehat{s(v_k)},\ldots,s(v_q))$$
$$= (-1)^{k-j-1}s_{q-1}(v_0,\ldots,\hat{v}_k,\ldots,v_q)$$

Suppose now that z is a q-cycle of K, so $\partial(z) = 0$. By our lemma, $\partial s_q(z) = s_{q-1}\partial(z) = 0$, and we see that $s_q(z)$ is a q-cycle of L. Similarly, if $b \in B_q(K)$ then $b = \partial c$ for some element $c \in C_{q+1}(K)$. But $\partial s_{q+1}(c) = s_q\partial(c) = s_q(b)$, giving $s_q(b) \in B_q(L)$. Therefore $s_q(Z_q(K)) \subseteq Z_q(L)$ and $s_q(B_q(K)) \subseteq B_q(L)$ as required.

We end this section with a little terminology which will considerably simplify the exposition in later sections. The collection of groups and homomorphisms

$$\ldots \xrightarrow{\partial} C_q(K) \xrightarrow{\partial} C_{q-1}(K) \xrightarrow{\partial} \ldots \xrightarrow{\partial} C_0(K) \xrightarrow{\partial} 0$$

will be referred to as the *chain complex* of K and written $C(K)$. Whenever we have a homomorphism $\phi_q : C_q(K) \to C_q(L)$ for each q satisfying

$$\partial \phi_q = \phi_{q-1}\partial$$

we abbreviate the whole collection to $\phi : C(K) \to C(L)$ and call ϕ a *chain map*.

So a simplicial map from K to L induces a chain map from the chain complex of K to that of L. The important property of a chain map is that it induces homomorphisms $\phi_{q*} : H_q(K) \to H_q(L)$ of homology groups. The proof is precisely the same as that given above for the special case where the chain map is induced by a simplicial map.

We shall often abbreviate our notation even further and simply write our homomorphisms as

$$\phi : C_q(K) \to C_q(L)$$
$$\phi_* : H_q(K) \to H_q(L)$$

when no confusion can arise from doing so.

(8.5) Lemma. *If $\psi : C(L) \to C(M)$ is a second chain map then $\psi \circ \phi : C(K) \to C(M)$ is a chain map and $(\psi \circ \phi)_* = \psi_* \circ \phi_* : H_q(K) \to H_q(M)$.*

The proof is left to the reader (Problem 20).

8.5 Stellar subdivision

Our object in this section is to show that barycentric subdivision does not change the homology groups of a complex. To this end, we shall explain how to barycentrically subdivide a complex by repeated application of a very simple operation called stellar subdivision.

Let K be a complex, A a simplex of K, and let v denote the barycentre of A. We chop up the simplexes of K as follows. Those simplexes which do not have A as a face are left untouched. If $A < B$, let L denote the subcomplex of the boundary of B consisting of those simplexes which *do not* have A as a face, and replace B by the cone with base L and apex v as in Fig. 8.6. This makes sense

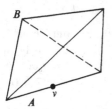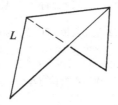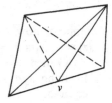

Figure 8.6   Cone on L with apex v

because adding v to the set of vertices of any simplex of L gives a collection of points which are in general position. We denote the resulting complex by K', and say that K' is formed from K by *stellar subdivision* of the simplex A.

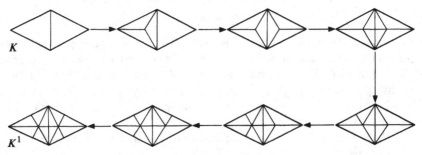

Figure 8.7

Suppose now we begin with a complex K and stellar-subdivide each of its simplexes, taking the simplexes in order of decreasing dimension. (The actual order inside any particular dimension does not matter.) Then we obtain the first barycentric subdivision as Fig. 8.7 indicates. And of course we may repeat the process, eventually producing any prescribed K^m.

(8.6) Theorem. *If K' is obtained from K by a single stellar subdivision, then K' and K have isomorphic homology groups.*

(8.7) Corollary. *Barycentric subdivision does not change the homology groups of a complex.*

We shall construct a chain map $\chi : C(K) \to C(K')$ and show that it induces iso-morphisms of homology groups. As usual, we need only specify the effect of χ on a typical oriented q-simplex σ of K so long as we are careful that $\chi(-\sigma) =$

$-\chi(\sigma)$. Suppose K' is obtained from K by stellar subdivision of the simplex A. If A is a face of σ, then σ is broken up into smaller q-simplexes when we form K'. We define $\chi(\sigma)$ to be the q-chain of K' which is the sum of those q-simplexes of K' that make up σ, each taken with the orientation induced from the given

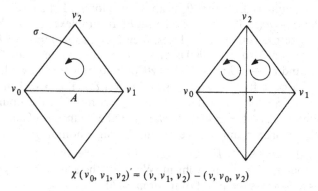

$$\chi(v_0, v_1, v_2)' = (v, v_1, v_2) - (v, v_0, v_2)$$

Figure 8.8

orientation of σ. Fig. 8.8 illustrates this definition. Put in a more formal way, if $\sigma = (v_0,\ldots,v_k, v_{k+1},\ldots,v_q)$, and if $v_0,\ldots,v_k$ are the vertices of A, then

$$\chi(\sigma) = \sum_{i=0}^{k} (-1)^i (v,v_0,\ldots,\hat{v}_i,\ldots,v_k,v_{k+1},\ldots,v_q).$$

If σ does not have A as a face we set $\chi(\sigma) = \sigma$.

(8.8) Lemma. χ *is a chain map.*

The proof involves computing the effect of $\partial\chi_q$ and $\chi_{q-1}\partial$ on a typical oriented q-simplex of K and showing that the answer is the same in both cases. We ask the reader to do this for himself. While the proof of a lemma like lemma (8.8) is, of necessity, computational, the geometry of the situation always tells us why the result ought to be true. When we apply χ to an oriented simplex of K, we may well chop it up as a sum of oriented simplexes of K', but the point is that any extra boundary created in this way cancels out. We see this very clearly in Fig. 8.8 where

$$\partial\chi_2(v_0, v_1, v_2) = \partial(v, v_1, v_2) - \partial(v, v_0, v_2)$$

$$= \chi_1\partial(v_0, v_1, v_2) - (v,v_2) + (v,v_2)$$

Not surprisingly we shall call χ the *subdivision chain map*. We now have homomorphisms $\chi_* : H_q(K) \to H_q(K')$ and we shall show that they are *isomorphisms*, thereby proving theorem (8.6).

Again let $v_0,\ldots,v_k$ denote the vertices of A, and v stand for its barycentre. Let θ be the simplicial map from K' to K which sends v to v_0 and which fixes all

187

the other vertices of K'. We use the same symbol θ for the induced chain map from $C(K')$ to $C(K)$. Now $\theta\chi$ is the identity homomorphism of $C_q(K)$ for each q, allowing us to conclude from lemma (8.5) that $H_q(K) \xrightarrow{\chi_*} H_q(K') \xrightarrow{\theta_*} H_q(K)$ is the identity.

We suspect, quite rightly, that θ_* is an inverse for χ_*. Let z be a q-cycle of K' and consider $z - \chi\theta(z)$. If L denotes the set of all simplexes of K' which have v as a vertex, together with all their faces, then L is a subcomplex of K' and is a *cone* with apex at v. Also, $z - \chi\theta(z)$ is a q-cycle of L since χ and θ are the identity outside of L and $\partial(z - \chi\theta(z)) = \partial(z) - \chi\theta\partial(z) = 0$. But we know all about the homology of a cone from Example 4 of Section 8.3: if $q > 0$ then $H_q(L) = 0$ and $H_0(L) \cong \mathbb{Z}$. So for $q > 0$ the cycle $z - \chi\theta(z)$ must be the boundary of a $(q + 1)$-chain of L, and therefore automatically the boundary of a $(q + 1)$-chain of K'. In other words, z and $\chi\theta(z)$ represent the same homology class in $H_q(K')$. This proves that $H_q(K') \xrightarrow{\theta_*} H_q(K) \xrightarrow{\chi_*} H_q(K')$ is the identity and completes our verification that χ_* is an isomorphism. We leave the special case $q = 0$ to the reader. This completes the proof of theorem (8.6).

If K^m is a barycentric subdivision of K, then we can produce it from K by a finite sequence of stellar subdivisions. The composition of all the associated subdivision chain maps gives a chain map $\chi : C(K) \to C(K^m)$ which we shall also refer to as a *subdivision chain map*. Going in the other direction, we have a simplicial map θ corresponding to each stellar subdivision: it is not unique but we agree to make a particular choice at each stage. The composition of all these will be denoted by the same symbol, so we write $\theta : |K^m| \to |K|$, and a map constructed in this way will be called a *standard simplicial map*.

Problems

20. Prove lemma (8.5).

21. Check that the subdivision map $\chi : C(K) \to C(K')$ is a chain map.

22. Give a second proof that barycentric subdivision does not change the Euler characteristic of a complex by showing that a single stellar subdivision does not change it.

23. If $s : |K^m| \to |L|$ simplicially approximates $f : |K^m| \to |L|$, if $n \geqslant m$, and if $\theta : |K^n| \to |K^m|$ is a standard simplicial map, prove that $s\theta : |K^n| \to |L|$ simplicially approximates $f : |K^n| \to |L|$.

8.6 Invariance

The homology groups of a complex, though defined using the simplicial structure of the complex, are invariants of the homotopy type of its underlying polyhedron. We shall now explain why this is the case. Some of the more computational details of our argument, which only cloud the issue on first reading, will be relegated to the problems at the end of the section.

The main theorems are these:

(8.9) Theorem. *Any map* $f:|K| \to |L|$ *induces a homomorphism* $f_*:H_q(K) \to H_q(L)$ *in each dimension.*†‡

(8.10)'Theorem. *If* f *is the identity map of* $|K|$ *then each* $f_*:H_q(K) \to H_q(K)$ *is the identity homomorphism, and if we have two maps* $|K| \xrightarrow{f} |L| \xrightarrow{g} |M|$ *then* $(g \circ f)_* = g_* \circ f_*:H_q(K) \to H_q(M)$ *for all q.*

(8.11) Theorem. *If* f,g$:|K| \to |L|$ *are homotopic maps then* $f_* = g_*:H_q(K) \to H_q(L)$ *for all q.*

It follows at once that if the polyhedra $|K|$ and $|L|$ have the same homotopy type, then K and L have isomorphic homology groups. For if $f:|K| \to |L|$ is a homotopy equivalence, with homotopy inverse g, then the composite homomorphisms

$$H_q(K) \xrightarrow{f_*} H_q(L) \xrightarrow{g_*} H_q(K)$$

$$H_q(L) \xrightarrow{g_*} H_q(K) \xrightarrow{f_*} H_q(L)$$

are both identity homomorphisms. Therefore $f_*:H_q(K) \to H_q(L)$ is an isomorphism for each q.

So if X is a compact triangulable space, we can choose a triangulation $t:|K| \to X$ and use it to define the homology groups $H_q(X)$ of X by $H_q(X) = H_q(K)$. *It does not matter which triangulation we choose, we shall always get the same groups (up to isomorphism).*

We have already seen how a simplicial map induces homomorphisms of homology groups. Not surprisingly, it is the simplicial approximation theorem (6.7) which allows us to pass to the general case of an arbitrary map. Let $f:|K| \to |L|$ be continuous and choose a simplicial approximation $s:|K^m| \to |L|$. Let $\chi:C(K) \to C(K^m)$ be the subdivision chain map and define the homomorphism $f_*:H_q(K) \to H_q(L)$ induced by f to be the composition

$$H_q(K) \xrightarrow{\chi_*} H_q(K^m) \xrightarrow{s_*} H_q(L)$$

Unfortunately, there is a choice involved in this definition, namely the choice of the simplicial approximation s. In order to show that this choice does not really matter, and in order to check theorems (8.10) and (8.11), we shall need the following two results:

1. *If* s,t$:|K| \to |L|$ *are 'close' simplicial maps, in the sense that for each simplex*

---

† We should really use the more cumbersome notation $f_{q*}:H_q(K) \to H_q(L)$.

‡ Remember that all simplicial complexes in this chapter are *finite*.

A *of* K *we can find a simplex* B *in* L *such that both* s(A) *and* t(A) *are faces of* B, *then* $s_* = t_* : H_q(K) \to H_q(L)$ *for all* q.

2. *If* f,g:$|K| \to |L|$ *are homotopic maps we can find a barycentric subdivision* K^m *and a sequence of simplicial maps* $s_1, \ldots, s_n : |K^m| \to |L|$ *such that* s_1 *simplicially approximates* f, s_n *simplicially approximates* g, *and each pair* s_i, s_{i+1} *are close in the sense of result* 1 *above.*

The proofs of 1, and 2 are broken up into easy stages in Problems 24–32 at the end of this section.

Suppose then that we simplicially approximate a given map $f:|K| \to |L|$ in two different ways via $s:|K^m| \to |L|$ and $t:|K^n| \to |L|$ where $n \geqslant m$. Let $\chi_1 : C(K) \to C(K^m)$, $\chi_2 : C(K^m) \to C(K^n)$ be subdivision chain maps, and let $\theta : |K^n| \to |K^m|$ be a standard simplicial map. If we want to show that we can use either s or t to define f_* we must check that

$$s_* \chi_{1*} = t_* \chi_{2*} \chi_{1*} : H_q(K) \to H_q(L)$$

It is easy to see that $s\theta : |K^n| \to |L|$ simplicially approximates $f:|K^n| \to |L|$. But so does t. Therefore $s\theta$ and t must be close simplicial maps and $s_* \theta_* = t_* : H_q(K^n) \to H_q(L)$. Since we also know that θ_* and χ_{2*} are inverse to one another, we have $t_* \chi_{2*} \chi_{1*} = s_* \theta_* \chi_{2*} \chi_{1*} = s_* \chi_{1*}$ as required. We now really do have a well-defined homomorphism $f_* : H_q(K) \to H_q(L)$, and we have proved theorem (8.9), the first of our three main theorems.

Proof of theorem (8.10). The first part of the theorem clearly follows by construction. Suppose we are given maps $|K| \xrightarrow{f} |L| \xrightarrow{g} |M|$. Choose a simplicial approximation $t:|L^n| \to |M|$ for $g:|L| \to |M|$, then a simplicial approximation $s:|K^m| \to |L^n|$ for $f:|K^m| \to |L^n|$. Let $\chi_1 : C(K) \to C(K^m)$, $\chi_2 : C(L) \to C(L^n)$ be subdivision chain maps, and let $\theta : |L^n| \to |L|$ be a standard simplicial map. We now have the following diagram of homology groups and homomorphisms:

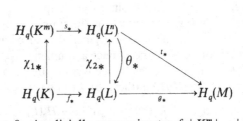

One easily checks that θs simplicially approximates $f:|K^m| \to |L|$, and that ts simplicially approximates $gf:|K^m| \to |M|$. Therefore

$$g_* \circ f_* = t_* \chi_{2*} \theta_* s_* \chi_{1*}$$
$$= t_* s_* \chi_{1*}$$
$$= (ts)_* \chi_{1*}$$
$$= (g \circ f)_*$$

as required.

Proof of theorem (8.11). This follows directly from results 1 and 2 above since, with the notation established in result 2,

$$f_* = s_{1*}\chi_* = s_{2*}\chi_* = \ldots = s_{n*}\chi_* = g_*$$

Having completed our invariance proofs, we can begin to solve some interesting problems. Referring back to the calculations in Section 8.3, we know that the homology groups of the n-sphere, $n > 0$, are as follows.

$$H_0(S^n) \cong \mathbb{Z}$$

$$H_n(S^n) \cong \mathbb{Z}$$

$$H_q(S^n) = 0 \qquad \text{for } q \neq 0, n$$

Also $H_0(S^0) \cong \mathbb{Z} \oplus \mathbb{Z}$ and $H_q(S^0) = 0$ for $q \neq 0$.

(8.12) Theorem. *If* m $\neq$ n *then* S^m *and* S^n *are not of the same homotopy type.*

Proof. $H_m(S^m)$ is isomorphic to $H_m(S^n)$ only when $m = n$.

(8.13) Corollary. *Two euclidean spaces are homeomorphic if and only if they have the same dimension.*

Proof. If $h: \mathbb{E}^m \to \mathbb{E}^n$ is a homeomorphism, then

$$S^{m-1} \simeq \mathbb{E}^m - \{0\} \cong \mathbb{E}^n - \{h(0)\} \simeq S^{n-1}$$

So by theorem (8.12) we must have $m = n$.

(8.14) Brouwer fixed-point theorem. *A map from* B^n *to itself must leave at least one point fixed.*

Proof. Mimic the proof given for the case $n = 2$ in Section 5.5, using the $(n-1)$th homology group in place of the fundamental group. (For an alternative proof, see theorem (9.18).)

(8.15) Theorem. *If* h: $|$K$| \to$ S *is a triangulation of a closed surface, then* S *is orientable if and only if the triangles of* K *can be oriented in a compatible manner.*

Proof. If S is orientable, we have already shown that the triangles of K can be compatibly oriented in Chapter 7. If S is not orientable, we can find a triangulation of S by a simplicial complex L whose simplexes cannot be compatibly oriented. Using L to calculate the homology of S gives $H_2(S) = 0$. But if we calculate using K, we must obtain the same answer. Therefore $H_2(K) = 0$, showing that the simplexes of K cannot be compatibly oriented.

Problems

24. If $s,t: |K^m| \to |L|$ both simplicially approximate $f: |K^m| \to |L|$, show that s and t are close simplicial maps.

25. Suppose $s,t: |K| \to |L|$ are simplicial, and assume we have a homomorphism $d_q: C_q(K) \to C_{q+1}(L)$, for each q, such that

$$d_{q-1}\partial + \partial d_q = t - s: C_q(K) \to C_q(L).$$

Show that s and t induce the same homomorphisms of homology groups. The collection of homomorphisms $\{d_q\}$ is called a *chain homotopy* between s and t.

In the next three problems we shall construct a chain homotopy between two close simplicial maps $s,t: |K| \to L|$. First a little terminology. If σ is an oriented simplex of K, call the smallest simplex of L which has both $s(\sigma)$ and $t(\sigma)$ as faces, the *carrier* of σ.

26. Given $\sigma = v \in C_0(K)$, define $d_0(\sigma) = 0$ if $s(v) = t(v)$, and $d_0(\sigma) = (s(v), t(v))$ if $s(v) \neq t(v)$. Check that $\partial d_0 = t - s: C_0(K) \to C_0(L)$ and that $d_0(\sigma)$ is a chain which lies in the carrier of σ. Where have you used the fact that s and t are close?

27. Suppose we have defined homomorphisms $d_i: C_i(K) \to C_{i+1}(L)$ for $0 \leqslant i \leqslant q - 1$ so that:
(a) $d_{i-1}\partial + \partial d_i = t - s: C_i(K) \to C_i(L)$;
(b) $d_i(\sigma)$ is always a chain in the carrier of σ.
If σ is an oriented q-simplex of K, prove that

$$\partial(t(\sigma) - s(\sigma) - d_{q-1}\partial(\sigma)) = 0$$

and deduce that

$$t(\sigma) - s(\sigma) - d_{q-1}\partial(\sigma) = c$$

for some chain $c \in C_{q+1}(L)$. The point is that the carrier of σ is a *cone*.

28. Set $d_q(\sigma) = c$ and show that you have completed an inductive construction for a chain homotopy between s and t.

29. You should now be able to show that close simplicial maps induce the same homomorphisms of homology groups.

30. Let $f,g: |K| \to |L|$ be maps, and write $d(f,g) < \delta$ if for any $x \in |K|$ the distance between $f(x)$ and $g(x)$ is less than δ. If δ is a Lebesgue number for the open covering of $|L|$ by the open stars of its vertices, and if $d(f,g) < \delta/3$, show that the sets

$$f^{-1}(\text{star }(v,L)) \cap g^{-1}(\text{star }(v,L)), \ v \text{ a vertex of } L,$$

form an open covering of $|K|$.

31. Use the conclusion of Problem 30 to find an integer m and a simplicial map $s: |K^m| \to |L|$ which simplicially approximates both $f: |K^m| \to |L|$ and $g: |K^m| \to |L|$.

32. Suppose $f,g : |K| \to |L|$ are homotopic maps, let $F : |K| \times I \to |L|$ be a specific homotopy between them, and write $f_t(x) = F(x,t)$. Given $\delta > 0$, find a positive integer n such that

$$d(f_{r/n}, f_{(r+1)/n}) < \delta,\ 0 \leqslant r < n.$$

Now verify result 2 of this section by finding, for each r, a common simplicial approximation to $f_{r/n}$ and $f_{(r+1)/n}$, provided n is large enough.

33. Give a second proof of corollary (8.13) using the fact that the one-point compactification of $\mathbb{E}^n$ is S^n.

34. Work through the details of the proof of theorem (8.14).

35. If two closed manifolds are homeomorphic, show they must have the same dimension.

36. An n-manifold with boundary consists of a second-countable Hausdorff space in which each point has a neighbourhood homeomorphic to either $\mathbb{E}^n$ or to the closed upper half-space $\mathbb{E}^n_+$. Those points which have a neighbourhood homeomorphic to $\mathbb{E}^n$ form the interior of the manifold. Those points x for which there is a neighbourhood U, and a homeomorphism $f : \mathbb{E}^n_+ \to U$ such that $f(0) = x$, form the boundary. Show that the interior and boundary of a manifold are disjoint. If $h : M \to N$ is a homeomorphism between two n-manifolds, prove that h induces a homeomorphism between the interior of M and the interior of N, and between the boundary of M and the boundary of N.

9. Degree and Lefschetz Number

9.1 Maps of spheres

The whole of this chapter will be devoted to applications of homology theory. We start by defining the 'degree' of a map from the n-sphere to itself, a concept due to Brouwer which allows one to decide whether or not two such maps are homotopic.

Choose a triangulation $h: |K| \to S^n$ of the n-sphere, and choose a generator $[z]$ for the infinite cyclic group $H_n(K)$. Given a map $f: S^n \to S^n$, we write f^h for the composite map $h^{-1}fh: |K| \to |K|$, and note that the induced homomorphism $f_*^h: H_n(K) \to H_n(K)$ sends $[z]$ to an integer multiple $\lambda[z]$ of itself. This integer λ is called the *degree of the map f* and is usually written $\deg f$.

The choices made above do not matter. For suppose we triangulate S^n via $t: |L| \to S^n$ and take $[w]$ as generator for $H_n(L)$. Then

$$f_*^t([w]) = (t^{-1}ft)_*([w])$$

$$= (\phi^{-1}f^h\phi)_*([w])$$

where ϕ is the homeomorphism $h^{-1}t: |L| \to |K|$. Remembering that f_*^h multiplies every element of $H_n(K)$ by λ, we have

$$f_*^t([w]) = \phi_*^{-1}(f_*^h(\phi_*([w])))$$

$$= \phi_*^{-1}(\lambda\phi_*([w]))$$

$$= \lambda[w]$$

In other words, f_*^t multiplies $[w]$ by the same integer λ, as required.

Homotopic maps have the same degree. For if $f \simeq g: S^n \to S^n$, then f^h and g^h are homotopic maps from $|K|$ to itself, and therefore induce the same homomorphism from $H_n(K)$ to $H_n(K)$. (It is in fact true that maps of the same degree are homotopic, though we shall not prove this here.) We also note that $\deg f \circ g = \deg f \times \deg g$ for any two maps $f,g: S^n \to S^n$. This is true because $(f \circ g)^h = f^h \circ g^h$, giving $(f \circ g)_*^h = f_*^h \circ g_*^h$ by theorem (8.10).

Clearly the degree of a homeomorphism must be ± 1; that of the identity map is $+1$; and that of a constant map (one which identifies all of S^n to a single point) is zero. We deduce at once that the identity map of S^n is never homotopic to a constant map.

Since we can work with any triangulation of S^n, we may as well construct a convenient one and stick to it from now on. Let v_i denote the point of $\mathbb{E}^{n+1}$ whose ith coordinate is 1 and all of whose other coordinates are zero, and let v_{-i} denote its antipode. Any collection $v_{i_1}, v_{i_2}, \ldots, v_{i_s}$ of such points for which $|i_1| < |i_2| < \ldots < |i_s|$ is in general position and therefore spans a simplex in $\mathbb{E}^{n+1}$. The collection of all these simplexes forms a simplicial complex which we shall denote by Σ (see Fig. 9.1 for the case $n = 2$). The polyhedron of Σ is nothing more than the set of points $(x_1, \ldots, x_{n+1}) \in \mathbb{E}^{n+1}$ which satisfy $\sum_{i=1}^{n+1} |x_i| = 1$, and radial projection $\pi : |\Sigma| \to S^n$ gives us a triangulation of S^n.

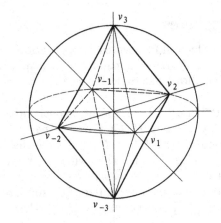

Figure 9.1

From now on we identify $H_n(S^n)$ with $H_n(\Sigma) = Z_n(\Sigma)$, and we specify a generator for this group as follows. Start with the simplex spanned by the vertices $v_1, v_2, \ldots, v_{n+1}$ and orient it as $\sigma = (v_1, v_2, \ldots, v_{n+1})$ then go round the complex orienting all the other top-dimensional simplexes in a compatible manner. The sum of the n-simplexes of Σ oriented in this way is an n-cycle z, which we know generates $Z_n(\Sigma)$.

Here is a second way of thinking of deg f. With $\pi : |\Sigma| \to S^n$ as above, choose a simplicial approximation $s : |\Sigma^m| \to |\Sigma|$ to f^π and orient all the top-dimensional simplexes of Σ^m by taking the orientations induced from those of the simplexes of Σ. In other words, orient each n-simplex of Σ^m exactly as it appears in $\chi(z)$, where $\chi : C(\Sigma) \to C(\Sigma^m)$ is the subdivision chain map. Let α denote the number of oriented n-simplexes τ of Σ^m such that $s(\tau) = \sigma$, and let β be the number such that $s(\tau) = -\sigma$.

(9.1) Theorem. deg $f = \alpha - \beta$.

Proof. The homomorphism $f_*^\pi : H_n(\Sigma) \to H_n(\Sigma)$ is by definition the composition $H_n(\Sigma) \xrightarrow{\chi_*} H_n(\Sigma^m) \xrightarrow{s_*} H_n(\Sigma)$, where χ is the subdivision chain map. Since there are

196

no simplexes of dimension greater than n, the homology groups $H_n(\Sigma)$, $H_n(\Sigma^m)$ are the same as the groups of n-cycles, and we can rewrite this as $Z_n(\Sigma) \xrightarrow{\chi_*} Z_n(\Sigma^m) \xrightarrow{s_*} Z_n(\Sigma)$. Now $\chi_*(z)$ is just the sum of all the oriented n-simplexes of Σ^m, and by the way s_* is defined the coefficient of the oriented simplex σ in $s_* \chi_*(z)$ is just $\alpha - \beta$. But $f_*^\pi(z) = s_* \chi_*(z) = (\deg f)z$, so $\deg f = \alpha - \beta$ as required.

(9.2) Theorem. *The antipodal map of S^n has degree* $(-1)^{n+1}$.

Proof. If f is the antipodal map of S^n then $f(v_i) = v_{-i}$ for each i and f^π is a simplicial homeomorphism. Recall that all the top-dimensional simplexes of Σ are oriented compatibly with $\sigma = (v_1, v_2, \ldots, v_{n+1})$. This means that the n-simplex obtained when we change v_1 to v_{-1} must be oriented as $-(v_{-1}, v_2, \ldots, v_{n+1})$ since it has to induce the opposite orientation to that induced by σ on the face spanned by $v_2, \ldots, v_{n+1}$. If we now change v_2 to v_{-2} in this new simplex, the resulting oriented simplex must be $(v_{-1}, v_{-2}, v_3, \ldots, v_{n+1})$, and so on. By interchanging all the v_i with their antipodes, one by one, we arrive at the oriented simplex $(-1)^{n+1}(v_{-1}, v_{-2}, \ldots, v_{-(n+1)})$. This obviously maps to $(-1)^{n+1}\sigma$ under f_*^π, and nothing else maps to $\pm \sigma$. Therefore $\deg f = (-1)^{n+1}$.

(9.3) Corollary. *A map from the n-sphere to itself which has no fixed points must have degree* $(-1)^{n+1}$.

Proof. If $f : S^n \to S^n$ has no fixed points, it is homotopic to the antipodal map via the homotopy $F : S^n \times I \to S^n$ given by

$$F(\mathbf{x}, t) = \frac{(1-t)f(\mathbf{x}) - t\mathbf{x}}{\|(1-t)f(\mathbf{x}) - t\mathbf{x}\|}$$

Therefore f has the same degree as the antipodal map.

(9.4) Corollary. *If n is even, and if* $f : S^n \to S^n$ *is homotopic to the identity, then* f *has a fixed point.*

Proof. Any map homotopic to the identity has degree $+1$, and by corollary (9.3) a map without fixed points should have degree $(-1)^{n+1} = -1$.

Given a group G acting as a group of homeomorphisms of a space X, we shall say that G acts *freely* if the only group element which has any fixed points is the identity element. Suppose now that G acts freely on S^n and that n is even. If $g, h \in G - \{e\}$ then $\deg g = \deg h = (-1)^{n+1} = -1$, therefore $\deg gh = +1$. But this means that gh must have a fixed point. By assumption, the given action is free, so $gh = e$, in other words $h = g^{-1}$. We have therefore proved the following result:

(9.5) Theorem. *Only* $\mathbb{Z}_2$ *(and the trivial group) can act freely on S^n when n is even.*

We know from the discussion of Lens spaces in Chapter 4 that any finite cyclic group can act freely on S^3. It is not hard to produce the same type of actions on the odd-dimensional spheres of higher dimension.

If for each point $\mathbf{x}$ of S^n we are given a vector in $\mathbb{E}^{n+1}$ which begins at $\mathbf{x}$, is tangent to S^n at $\mathbf{x}$, and whose endpoint $v(\mathbf{x})$ varies continuously in $\mathbb{E}^{n+1}$ as $\mathbf{x}$ varies in S^n, then we say that we have a *continuous vector field* on S^n. If in addition $v(\mathbf{x})$ is never equal to $\mathbf{x}$, we say we have a *nonvanishing* field. When n is odd, it is easy to construct a nonvanishing vector field on S^n. For suppose $n = 2m - 1$, let $\mathbf{x} = (x_1, \ldots, x_{2m})$ be a point of S^n, and observe that the vector represented by $(-x_{m+1}, \ldots, -x_{2m}, x_1, \ldots, x_m)$ is orthogonal to the radius vector through $\mathbf{x}$. We now assign to $\mathbf{x}$ the vector which begins at $\mathbf{x}$ and ends at the point denoted by
$$v(\mathbf{x}) = (x_1 - x_{m+1}, \ldots, x_m - x_{2m}, x_{m+1} + x_1, \ldots, x_{2m} + x_m).$$

For n even, no such field can be found. For the map $f : S^n \to S^n$ defined by $f(x) = v(\mathbf{x})/\| v(\mathbf{x}) \|$ is clearly homotopic to the identity and so must have a fixed point by corollary (9.4). In other words, the vector field must vanish at some point x of S^n. We have proved the following result:

(9.6) Theorem. *S^n admits a continuous nonvanishing vector field if and only if n is odd.*

The lack of any continuous nonvanishing vector field on S^2 is a favourite result and is nicknamed the 'hairy ball theorem'. If we have a hair growing out from each point on the surface of a ball, any attempt to comb the hairs smoothly round the ball meets with defeat. Just about the best we can do if we want the hair to lie down smoothly is to comb the ball as shown in Fig. 9.2, leaving the odd bald spot. If we could comb the hair smooth then the tangent vectors to the hairs would contradict theorem (9.6) for $n = 2$.

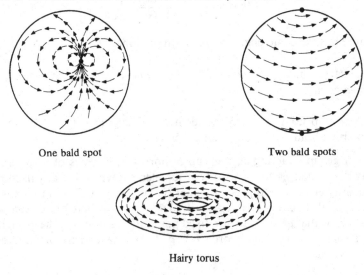

One bald spot   Two bald spots

Hairy torus

Figure 9.2

We can, however, comb the hairy torus (Fig. 9.2). In fact, the torus is the only orientable hairy surface which can be combed smooth, as we shall see in Section 9.4.

Problems

1. The map $z \mapsto z^n$ from the complex plane to itself extends in a unique way to a map from S^2 to S^2. What is the degree of this map?

2. Prove that the set of homotopy classes of maps from S^n to itself is infinite for $n \geqslant 1$, by constructing a map of degree k for each integer k.

3. If the degree of $f : S^n \to S^n$ is not $+1$, show that f must map some point to its antipode.

4. Show that the antipodal map of the circle is homotopic to the identity.

5. Let X and Y be subsets of $\mathbb{E}^n$ positioned in such a way that if x_1, x_2 are distinct points of X, and y_1, y_2 are distinct points of Y, then the line segments which join x_1 to y_1 and x_2 to y_2 do not intersect. Write $X * Y$ for the union of all the line segments which join a point of X to a point of Y, and call this the *join* of X and Y. Show that a typical point of the join can be written as $tx + (1 - t)y$, where $x \in X$, $y \in Y$, and $0 \leqslant t \leqslant 1$, and that this representation is unique provided the point in question does not lie in X or in Y.

6. If $X \cong [0,1] \cong Y$ show that $X * Y$ is a tetrahedron. More generally, if X is an m-simplex and Y an n-simplex, show that $X * Y$ is an $(m + n + 1)$-simplex. Deduce that $B^m * B^n \cong B^{m+n+1}$ and $S^m * S^n \cong S^{m+n+1}$.

7. Suppose we have two joins $X * Y$ and $X' * Y'$. Invent a definition of the join $f * g : X * Y \to X' * Y'$ of two maps $f : X \to X'$, $g : Y \to Y'$. Show that if both f and g are homotopic to the identity, then so is $f * g$.

8. Prove that an odd-dimensional sphere is a join of circles, then show that the antipodal map of an odd-dimensional sphere is homotopic to the identity.

9. Given maps $f : S^m \to S^m$, $g : S^n \to S^n$ show that $\deg f * g = (\deg f).(\deg g)$.

10. If $f : S^n \to S^n$ is a map, and if n is even, show that f^2 must have a fixed point. Even better, prove that either f has a fixed point, or it sends some point to its antipode.

9.2 The Euler–Poincaré formula

Recall that the rank of the free abelian part of $H_q(K)$ is called the qth Betti number of K and written β_q. We will prove:

(9.7) Euler–Poincaré formula. *The Euler characteristic of a finite complex* K *is given by the formula*

$$\chi(K) = \sum_{q=0}^{n} (-1)^q \beta_q$$

where n *is the dimension of* K.

Since the homology groups $H_q(K)$, and therefore the numbers β_q, depend only on the homotopy type of $|K|$ we at once deduce the same property for $\chi(K)$:

(9.8) Corollary. *Complexes whose polyhedra are homotopy equivalent have the same Euler characteristic.*

A special case of this result was widely advertised in Chapter 1 and provided much of the impetus for all the machinery we have developed since then. There we worked with rather concrete 'polyhedra' made up of plane polygonal faces fitting together nicely, and we claimed as theorem (1.2) that if two such were topologically equivalent they had the same Euler number (defined as vertices minus edges plus faces). But we can subdivide such a 'polyhedron' into a two-dimensional simplicial complex simply by chopping up each face as a cone with apex its centre of gravity. The Euler characteristic of the resulting complex is precisely the Euler number of the original 'polyhedron', allowing us to deduce theorem (1.2) from corollary (9.8).

In order to prove theorem (9.7), it is convenient to reinterpret the Betti numbers slightly as follows. Suppose we go through the process of setting up the homology groups of a complex, but allowing *rational numbers* as coefficients when we form linear combinations of oriented simplexes. To be precise, consider formal linear combinations $r_1\sigma_1 + \ldots + r_s\sigma_s$, where each σ_i is an oriented q-simplex of K and each r_i is a rational number. Clearly the set of all such expressions forms a vector space V over the rational field $\mathbb{Q}$ in a natural way. Let W be the subspace of V spanned by elements of the form $\sigma + \tau$, where σ, τ are the same q-simplex with opposite orientations. We call the quotient space V/W the vector space of rational q-chains of K, and denote it by $C_q(K,\mathbb{Q})$. The dimension of $C_q(K,\mathbb{Q})$ over $\mathbb{Q}$ is just the number of q-simplexes in K. We can produce a boundary homomorphism and use it to define rational q-cycles, and rational bounding cycles, exactly as before. In this setting, the boundary homomorphism is a linear map of vector spaces over $\mathbb{Q}$, so the rational q-cycles $Z_q(K,\mathbb{Q})$, and the bounding cycles $B_q(K,\mathbb{Q}) \subseteq Z_q(K,\mathbb{Q})$, form subspaces of $C_q(K,\mathbb{Q})$. The quotient space $H_q(K,\mathbb{Q}) = Z_q(K,\mathbb{Q})/B_q(K,\mathbb{Q})$ is called the *q*th homology group of K with rational coefficients.

(9.9) Lemma. β_q *is the dimension of* $H_q(K,\mathbb{Q})$ *as a vector space over* $\mathbb{Q}$.

Proof. Choose a minimal set of generators $[z_1], \ldots, [z_{\beta_q}], [w_1], \ldots, [w_{\gamma_q}]$ for $H_q(K)$, where the $[z_i]$ generate the free part of the group and the $[w_i]$ all have finite order. A q-cycle z with integer coefficients can be thought of as having rational coefficients, and it therefore determines an element of $H_q(K,\mathbb{Q})$ which we denote by $\{z\}$ in order to distinguish it from the corresponding element $[z]$ of $H_q(K)$. Suppose

$$\frac{a_1}{b_1}\sigma_1 + \ldots + \frac{a_s}{b_s}\sigma_s$$

is a rational q-cycle, the a_i, b_i being integers. Then

$$\frac{a_1}{b_1}\sigma_1 + \ldots + \frac{a_s}{b_s}\sigma_s = \frac{1}{b_1 b_2 \ldots b_s} \times \text{(a cycle with integer coefficients)}$$

$$= \frac{1}{b_1 b_2 \ldots b_s} \times \text{(a linear combination of } z_i\text{'s and } w_i\text{'s)}.$$

Therefore the elements $\{z_1\},\ldots,\{z_{\beta_q}\}, \{w_1\},\ldots,\{w_{\gamma_q}\}$ span $H_q(K,\mathbb{Q})$.

If $[w]$ is an element of $H_q(K)$ which has finite order m, then mw is the boundary of a $(q + 1)$-chain with integer coefficients. Dividing by m, we find that w is itself the boundary of a $(q + 1)$-chain which has rational coefficients, and consequently $\{w\} = 0$. Therefore $H_q(K,\mathbb{Q})$ is spanned by $\{z_1\},\ldots,\{z_{\beta_q}\}$.

Finally, if some linear combination of $z_1,\ldots,z_{\beta_q}$ with rational coefficients is the boundary of a rational $(q + 1)$-chain, multiplying by the product of the denominators of all the rational coefficients involved produces a linear combination of the z_i in which the coefficients are integers, and which is the boundary of a $(q + 1)$-chain with integer coefficients. But this can happen only if the coefficient of each z_i is zero. Therefore the original rational coefficients must all have been zero and we conclude that $\{z_1\},\ldots,\{z_{\beta_q}\}$ are linearly independent over $\mathbb{Q}$.

Proof of theorem (9.7). By definition, $\chi(K) = \sum_{q=0}^{n} (-1)^q \alpha_q$ where α_q is the number of q-simplexes in K. We shall abbreviate $C_q(K,\mathbb{Q})$ simply to C_q, and use corresponding abbreviations for the subspaces of cycles and bounding cycles. Choose bases for the C_q as follows. Since K has no $(n + 1)$-simplexes, $B_n = 0$, and therefore β_n is the dimension of Z_n. Begin by selecting a basis $z_1^n,\ldots,z_{\beta_n}^n$ for Z_n, then extending this by elements $c_1^n,\ldots c_{\gamma_n}^n$ to a basis for the whole of C_n. Applying ∂ to these basis elements gives us a basis $\partial c_1^n,\ldots,\partial c_{\gamma_n}^n$ for B_{n-1}, which we extend by $z_1^{n-1},\ldots z_{\beta_{n-1}}^{n-1}$ to a basis for Z_{n-1}, and then extend further by $c_1^{n-1},\ldots,c_{\gamma_{n-1}}^{n-1}$ to a basis for all of C_{n-1}. Note that the dimension of Z_{n-1} minus that of B_{n-1} is indeed β_{n-1} by lemma (9.9). Continue in this way. The general step is to use $\partial c_1^{q+1},\ldots,\partial c_{\gamma_{q+1}}^{q+1}$ as a basis for B_q, extend via $z_1^q,\ldots,z_{\beta_q}^q$ to Z_q, then via $c_1^q,\ldots,c_{\gamma_q}^q$ to C_q. The process terminates with the basis $\partial c_1^1,\ldots,\partial c_{\gamma_1}^1, z_1^0, \ldots,z_{\beta_0}^0$ for $Z_0 = C_0$.

Now α_q is the dimension of C_q and therefore equals $\gamma_{q+1} + \beta_q + \gamma_q$. Hence

$$\sum_{q=0}^{n} (-1)^q \alpha_q = \sum_{q=0}^{n} (-1)^q (\gamma_{q+1} + \beta_q + \gamma_q)$$

$$= \sum_{q=0}^{n} (-1)^q \beta_q$$

since each γ_q occurs with the sign $(-1)^q$ and with $(-1)^{q-1}$ if $0 < q \leqslant n$, and both of γ_0 and γ_{n+1} are zero.

Problems

11. Show that the Euler characteristic of the standard orientable surface of genus g is $2 - 2g$.

12. Show that the Euler characteristic of the standard nonorientable surface of genus g is $2 - g$.

13. Calculate the Euler characteristic of the sphere with k holes punched in.

14. Calculate the Euler characteristic of $H(p,r)$ and $M(q,s)$.

15. Let K and L be finite complexes. By triangulating $| K | \times | L |$ appropriately, show

$$\chi(| K | \times | L |) = \chi(| K |) \cdot \chi(| L |).$$

16. Use Problem 14 of Chapter 7 to work out the Euler characteristic of the Lens space $L(p,q)$. Now write down the Betti numbers of this space.

17. What is $\chi(P^n)$? What is $\chi(S^m \times S^n)$?

18. Show that the Euler characteristic of the n-dimensional torus $T^n = S^1 \times S^1 \times \ldots \times S^1$ is zero using Problem 15. Now give a second proof by finding a free simplicial action of Z_2 on T^n which has T^n as quotient space.

9.3   The Borsuk–Ulam theorem

In order to prove the topological invariance of the Euler characteristic, we 'changed coefficients' from the integers to the rational numbers. If we examine carefully the definition of the homology groups of a complex K, we see that it makes sense to replace the integers by any abelian group G. A q-chain now becomes a formal linear combination $g_1\sigma_1 + \ldots + g_s\sigma_s$ where the g_i belong to G, the σ_i are oriented q-simplexes of K, and $(-g)\sigma$ is always identified with $g(-\sigma)$. The remainder of the construction is automatic and it results in the so-called homology groups of K with coefficients in G. We do not have the space to work in this degree of generality here, but we would like to mention a second special case, namely the case of $\mathbb{Z}_2$ coefficients.

Consider linear combinations as above where each coefficient g_i is either 0 or 1, and agree to add these coefficients mod 2. The identifications $(-g)\sigma = g(-\sigma)$ reduce in this case to $\sigma = -\sigma$ (taking $g = 1$); in other words there is no longer any need to orient each simplex of K, we can simply work with linear combinations of *unoriented* q-simplexes of K in which each coefficient is either 0 or 1. Such linear combinations are called 'mod-2' q-chains of K, and they form a finitely generated abelian group $C_q(K,\mathbb{Z}_2)$ in which each element has order 2. Notice that every mod-2 q-chain has a geometrical interpretation because it is the sum of certain q-simplexes of K.

The mod-2 boundary of a q-simplex is just the sum of its $(q - 1)$-dimensional faces. Extending linearly to sums of simplexes, we have a boundary homomorphism

$$\partial : C_q(K, \mathbb{Z}_2) \to C_{q-1}(K, \mathbb{Z}_2)$$

which satisfies $\partial^2 = 0$. The kernel of this homomorphism divided by those elements which lie in the image of $\partial : C_{q+1}(K, \mathbb{Z}_2) \to C_q(K, \mathbb{Z}_2)$ is the q-th homology group of K with $\mathbb{Z}_2$ coefficients, and is written $H_q(K, \mathbb{Z}_2)$. Clearly, each element of this group has order 2, so $H_q(K, \mathbb{Z}_2)$ is a finite sum of copies of $\mathbb{Z}_2$. It is easy to redo the invariance proofs of Chapter 8 in this setting and show that these mod-2 homology groups depend only on the homotopy type of $|K|$. (In fact the homology groups of a complex with coefficients in an arbitrary abelian group G are completely determined by the integral homology groups of the complex.)

If we work with $\mathbb{Z}_2$ as coefficient group, then we do of course lose some information, because we are throwing away any consideration of orientation. We can see this clearly in the case of two surfaces like the torus and Klein bottle, which are nicely distinguished by their second homology groups with integer coefficients. However, when we use $\mathbb{Z}_2$ as coefficient group, the second homology group is $\mathbb{Z}_2$ in *both* cases, since we obtain a 2-cycle by taking the sum of all the triangles in any triangulation of the surface, and this is the only nonzero 2-cycle. (When we take the boundary of this sum, every edge of the triangulation occurs twice and therefore disappears since we are working mod 2.) We invite the reader to work out the mod-2 homology groups of each of the standard closed surfaces.

We shall use $\mathbb{Z}_2$ coefficients to give a reasonably efficient proof of the following result:

(9.10) Theorem. *Let* $f : S^n \to S^n$ *be a map which preserves antipodal points, in other words* $f(-x) = -f(x)$ *for every point* x *of* S^n. *Then* f *has odd degree.*

Let $\pi : |\Sigma| \to S^n$ be the triangulation described in Section 9.1 and, as before, write f^π for the map $\pi^{-1} f \pi : |\Sigma| \to |\Sigma|$.

(9.11) Lemma. *The map* f^π *admits a simplicial approximation* $s : |\Sigma^m| \to |\Sigma|$ *which preserves antipodal points.*

Proof. The proof is made easy by the amount of choice available in the construction of a simplicial approximation. Choose m large enough, as in theorem (6.7), so that for each vertex v of Σ^m we can find a vertex w of Σ for which

$$f^\pi(\text{star } (v, \Sigma^m)) \subseteq \text{star } (w, \Sigma) \tag{*}$$

Note that if $\phi : |\Sigma| \to |\Sigma|$ is the antipodal map, then $\phi f = f \phi$, giving

$f^{\pi}(\text{star}\,(\phi(v),\Sigma^m)) \subseteq \text{star}\,(\phi(w),\Sigma)$. Now select one half of the vertices of Σ^m in such a way that no two are antipodal, and for each such vertex v make a choice of w in Σ so that (*) is satisfied. Define $s(v) = w$ and complete the definition of s on the remaining vertices of Σ^m by $s(\phi(v)) = \phi(w)$. The first half of the proof of theorem (6.7) shows us that this mapping of vertices determines a simplicial approximation s to $f^{\pi}:|\Sigma^m| \to |\Sigma|$, and by construction s preserves antipodal points.

Proof of theorem (9.10). We know how to calculate the degree of f using a simplicial approximation by theorem (9.1). Now if α and β are integers, then $\alpha - \beta$ and $\alpha + \beta$ are even, or odd, together. Therefore in order to show f has odd degree, all we have to do is to verify that s maps an odd number of n-simplexes of Σ^m onto each n-simplex of Σ. We reinterpret this in terms of homology with mod-2 coefficients as follows. The sum of all the n-simplexes of Σ is the only nonzero mod-2 n-cycle, giving $H_n(\Sigma,\mathbb{Z}_2) \cong \mathbb{Z}_2$. Similarly $H_n(\Sigma^m,\mathbb{Z}_2) \cong \mathbb{Z}_2$, the nonzero element being the sum of the n-simplexes of Σ^m. Now s maps an odd number of n-simplexes of Σ^m onto each n-simplex of Σ if and only if s sends the unique nonzero mod-2 n-cycle of Σ^m to that of Σ; in other words, if and only if $s_*:H_n(\Sigma^m,\mathbb{Z}_2) \to H_n(\Sigma,\mathbb{Z}_2)$ is an isomorphism.

We need a little extra notation. Write Σ_k for the subcomplex of Σ made up of simplexes whose vertices are the points v_i, v_{-i} where $1 \leqslant i \leqslant k + 1$. So Σ_0 consists of the two points v_1, v_{-1} and Σ_{n-1} is the 'equator' in $\Sigma_n = \Sigma$. Let z_k be the sum of all the k-simplexes of Σ_k^m, and note that

$$z_k = c_k + \phi(c_k)$$

where a k-simplex of Σ_k^m lies in c_k if and only if the k-simplex of Σ_k which contains it has v_{k+1} as a vertex. Note also that $\partial(c_k) = z_{k-1}$.

Suppose $s(z_n)$ is the zero element of $Z_n(\Sigma,\mathbb{Z}_2) = H_n(\Sigma,\mathbb{Z}_2)$, then $s(c_n) + s\phi(c_n) = 0$. But s preserves antipodal points and therefore commutes with ϕ, giving $s(c_n) + \phi s(c_n) = 0$ or equivalently, since we are working mod 2, $s(c_n) = \phi s(c_n)$. If this is the case, $s(c_n)$ can be written as

$$s(c_n) = d_n + \phi(d_n)$$

where d_n is the sum of those n-simplexes in $s(c_n)$ which contain the vertex v_{n+1}. Taking the boundary of both sides, we now have

$$s\partial(c_n) = s(z_{n-1}) = s(c_{n-1}) + \phi s(c_{n-1})$$

$$= \partial(d_n) + \phi\partial(d_n)$$

and therefore

$$s(c_{n-1}) + \partial(d_n) = \phi(s(c_{n-1}) + \partial(d_n))$$

So we can write the $(n-1)$-chain $s(c_{n-1}) + \partial d_n$ of Σ in the form $s(c_{n-1}) + \partial d_n = d_{n-1} + \phi(d_{n-1})$, where d_{n-1} is the sum of those simplexes in the chain which contain v_{n+1}, and those which contain v_n but not $v_{-(n+1)}$.

Applying the boundary operator again, we obtain

$$s\partial(c_{n-1}) + \partial^2(d_n) = s(z_{n-2}) = s(c_{n-2}) + \phi s(c_{n-2})$$

$$= \partial(d_{n-1}) + \phi\partial(d_{n-1})$$

and we can keep repeating this process until we arrive at

$$s(z_0) = \partial(d_1) + \phi\partial(d_1)$$

where d_1 is a 1-chain of Σ. But this is impossible because $s(z_0)$ is a single pair of antipodal vertices of Σ, whereas $\partial(d_1) + \phi\partial(d_1)$ consists of an even number of such pairs of vertices. This contradiction proves that $s(z_n)$ is a nonzero mod-2 n-cycle of Σ, and therefore that s induces an isomorphism of $H_n(\Sigma^m, \mathbb{Z}_2)$ with $H_n(\Sigma, \mathbb{Z}_2)$ as required. This completes the proof of theorem (9.10).

The above result has some interesting consequences.

(9.12) Theorem. *If* $f : S^m \to S^n$ *sends antipodal points to antipodal points, then* $m \leqslant n$.

Proof. Suppose $m > n$ and let g denote the restriction of f to the n-sphere consisting of those points of S^m whose last $m - n$ coordinates are all zero. Then g is a map from S^n to S^n which preserves antipodal points, and should therefore have odd degree by theorem (9.10). But g is homotopic to a constant map because it extends over the $(n + 1)$-ball consisting of those points of S^m whose last $m - n - 1$ coordinates are all zero and whose $(m - n)$th coordinates are nonnegative. So the degree of g is zero and we have a contradiction.

(9.13) Borsuk–Ulam theorem. *Any map* $f : S^n \to \mathbb{E}^n$ *must identify a pair of antipodal points of* S^n.

Proof. Suppose $f(\mathbf{x})$ and $f(-\mathbf{x})$ are never equal. Then the formula

$$g(\mathbf{x}) = \frac{f(\mathbf{x}) - f(-\mathbf{x})}{\|f(\mathbf{x}) - f(-\mathbf{x})\|}$$

defines a map from S^n to S^{n-1} which preserves antipodal points, contradicting theorem (9.12).

(9.14) Corollary. *It is impossible to embed* S^n *in* $\mathbb{E}^n$.

Proof. S^n is not homeomorphic to a subset of $\mathbb{E}^n$, by theorem (9.13).

(9.15) Lusternik–Schnirelmann theorem. *If* S^n *is covered by* $n + 1$ *closed sets, then one of the sets contains a pair of antipodal points.*

Proof. Suppose $A_1, \ldots, A_{n+1}$ are closed subsets of S^n whose union is all of S^n. The function $f : S^n \to \mathbb{E}^n$ defined by $f(\mathbf{x}) = (d(\mathbf{x}, A_1), \ldots, d(\mathbf{x}, A_n))$, where $d(\mathbf{x}, A_i)$ is the distance of the point $\mathbf{x}$ from A_i, is continuous and must therefore identify a pair of antipodal points. In other words, we can find a point $\mathbf{y}$ of S^n with the property $d(\mathbf{y},A_i) = d(-\mathbf{y},A_i)$ for $1 \leqslant i \leqslant n$. If $d(\mathbf{y},A_i) > 0$ for $1 \leqslant i \leqslant n$, then $\mathbf{y}$ and $-\mathbf{y}$ lie in A_{n+1}, since $A_1,...,A_{n+1}$ cover S^n. On the other hand, if $d(\mathbf{y},A_i) = 0$ for some i, we have both $\mathbf{y}$ and $-\mathbf{y}$ in A_i because each A_i is a closed set.

Problems

19. Assume the Borsuk–Ulam theorem and give a proof of theorem (9.12).

20. Check that only n of the sets $A_1,...,A_{n+1}$ need be closed for the argument of theorem (9.15) to work.

21. If a map from S^n to S^n extends over B^{n+1}, show it must identify a pair of antipodal points of S^n. Prove that the same conclusion holds under the weaker assumption that f have even degree.

22. (*Ham sandwich theorem*). Let A_1, A_2, A_3 be bounded convex subsets of $\mathbb{E}^3$, and define a function $f:S^3 \to \mathbb{E}^3$ using them as follows. A point $x \in S^3$ determines a unique three-dimensional hyperplane $P(x)$ in $\mathbb{E}^4$ which is perpendicular to the radius vector through x and goes through the point $(0,0,0,\frac{1}{2})$. Let $f_i(x)$ be the volume of that part of A_i which lies on the same side of $P(x)$ as x, and define $f(x) = (f_1(x), f_2(x), f_3(x))$. Check the continuity of f, then find a plane in $\mathbb{E}^3$ which bisects each of A_1, A_2, A_3 by applying the Borsuk–Ulam theorem to f.

23. Work out the mod-2 homology groups of an arbitrary closed surface and compare them with the integral homology groups.

24. Define the qth mod-2 Betti number $\bar{\beta}_q$ of a finite complex K to be the number of copies of $\mathbb{Z}_2$ in $H_q(K,\mathbb{Z}_2)$. Show that

$$\sum_{q=0}^{n} (-1)^q \bar{\beta}_q = \chi(K)$$

where n is the dimension of K.

9.4 The Lefschetz fixed-point theorem

Let $f:X \to X$ be a map from a compact triangulable space to itself. Fix a triangulation $h:|K| \to X$ and let n denote the dimension of K. If we work with rational coefficients, the homology groups $H_q(K,\mathbb{Q})$ are all vector spaces over $\mathbb{Q}$, and the homomorphisms $f_{q*}^h:H_q(K,\mathbb{Q}) \to H_q(K,\mathbb{Q})$ are linear maps. The alternating sum of the traces of these linear maps, that is to say the number

$$\sum_{q=0}^{n} (-1)^q \operatorname{trace} f_{q*}^h$$

is called the *Lefschetz number of* f and written Λ_f. As usual, the choice of triangulation does not matter: any other triangulation will give the same value for Λ_f. We leave the reader to check this.

Since homotopic maps induce the same homomorphisms of homology, we see that $\Lambda_f = \Lambda_g$ whenever f is homotopic to g.

(9.16) Lefschetz fixed-point theorem. *If $\Lambda_f \neq 0$ then f has a fixed point.*

In order to understand the proof, we look at the simplest possible case, namely that where X is the polyhedron of a finite simplicial complex K and $f : |K| \to |K|$ is a simplicial map. Suppose f has no fixed points, then if A is a simplex of K we know that $f(A) \neq A$. Now orienting each q-simplex of K in some way gives a basis over $\mathbb{Q}$ for the vector space $C_q(K,\mathbb{Q})$; with respect to this basis, the matrix representing the linear map $f_q : C_q(K,\mathbb{Q}) \to C_q(K,\mathbb{Q})$ will have zeros along the diagonal, and therefore have trace zero. The crucial observation now (provided by theorem (9.17) below) is that whether we calculate the Lefschetz number of f at homology level, or at chain level, does not matter. In other words,

$$\sum_{q=0}^{n} (-1)^q \operatorname{trace} f_q = \sum_{q=0}^{n} (-1)^q \operatorname{trace} f_{q*}$$

giving $\Lambda_f = 0$. As usual, only technical difficulties are involved in passing from this special situation to the general case.

(9.17) **Hopf trace theorem.** *If K is a finite complex of dimension n, and $\phi : C(K,\mathbb{Q}) \to C(K,\mathbb{Q})$ a chain map, then*

$$\sum_{q=0}^{n} (-1)^q \operatorname{trace} \phi_q = \sum_{q=0}^{n} (-1)^q \operatorname{trace} \phi_{q*}.$$

Proof. Choose a 'standard' basis for $C(K,\mathbb{Q})$ as in the proof of theorem (9.7). The basis of $C_q(K,\mathbb{Q})$ therefore consists of elements

$$\partial c_1^{q+1}, \ldots, \partial c_{\gamma_{q+1}}^{q+1}, z_1^q, \ldots, z_{\beta_q}^q, c_1^q, \ldots, c_{\gamma_q}^q$$

A diagonal element of the matrix of ϕ_q with respect to this basis is obtained by taking a basis element w, expressing $\phi_q(w)$ in terms of the basis (i.e., as a linear combination of its elements), and reading off the coefficient of w. We shall call this coefficient $\lambda(w)$. With this convention the trace of ϕ_q is

$$\sum_{j=1}^{\gamma_{q+1}} \lambda(\partial c_j^{q+1}) + \sum_{j=1}^{\beta_q} \lambda(z_j^q) + \sum_{j=1}^{\gamma_q} \lambda(c_j^q)$$

But ϕ is a chain map, in other words $\phi\partial = \partial\phi$, giving $\lambda(\partial c_j^{q+1}) = \lambda(c_j^{q+1})$. Therefore

$$\sum_{q=0}^{n} (-1)^q \operatorname{trace} \phi_q = \sum_{q=0}^{n} (-1)^q \sum_{j=1}^{\beta_q} \lambda(z_j^q)$$

the other terms cancelling out in pairs. Since $\{z_1^q\},\ldots,\{z_{\beta_q}^q\}$ form a basis for the homology group $H_q(K,\mathbb{Q})$, we have

$$\sum_{j=1}^{\beta_q} \lambda(z_j^q) = \text{trace } \phi_{q*}$$

completing the argument.

Proof of theorem (9.16). We suppose that f, and therefore f^h, has no fixed points, and try to show $\Lambda_f = 0$. Let d be the metric on $|K|$ induced from the surrounding euclidean space. The real-valued function on $|K|$ given by $x \mapsto d(x,f^h(x))$ is never zero since f^h has no fixed points, and attains its lower bound $\delta > 0$ since $|K|$ is compact. By changing to a suitable barycentric subdivision if necessary, we may assume that the mesh of K is less than $\delta/3$.

Choose a simplicial approximation $s:|K^m| \to |K|$ to $f^h:|K^m| \to |K|$ and, as usual let $\chi:C(K,\mathbb{Q}) \to C(K^m,\mathbb{Q})$ denote the subdivision chain map. By definition f_{q*}^h is the composition

$$H_q(K,\mathbb{Q}) \xrightarrow{\chi_{q*}} H_q(K^m,\mathbb{Q}) \xrightarrow{s_{q*}} H_q(K\mathbb{Q})$$

Therefore, by the Hopf trace theorem, we can show that $\Lambda_f = 0$ by showing that each of the linear maps $s_q\chi_q:C_q(K,\mathbb{Q}) \to C_q(K,\mathbb{Q})$ has trace zero.

Let σ be an oriented q-simplex of K and let τ be an oriented q-simplex of K^m which lies in the chain $\chi_q(\sigma)$. So τ is contained in σ. If $x \in \tau$ we have $d(s(x),f^h(x)) < \delta/3$ since s simplicially approximates f^h; consequently we must have $d(x,s(x)) > 2\delta/3$. If now $y \in \sigma$, then $d(x,y) < \delta/3$ giving $d(y,s(x)) > \delta/3$. This means that $s(x)$ and y do not lie in the same simplex of K, and therefore $s(\tau) \neq \sigma$. So our simplex σ has coefficient zero in the chain $s_q\chi_q(\sigma)$, and trace $s_q\chi_q = 0$ as required.

As in Chapter 5, we shall say that a space X has the *fixed-point property* if every map from X to itself has a fixed point.

(9.18) Theorem. *A compact triangulable space which has the same rational homology groups as a point has the fixed-point property.*

Proof. If we take a triangulation $h:|K| \to X$ of our space and calculate the homology groups of K we are told the answer is $H_0(K,\mathbb{Q}) \cong \mathbb{Q}$ and $H_q(K,\mathbb{Q}) = 0$ otherwise. So for any map $f:X \to X$, the induced homomorphisms f_{q*}^h are all zero when $q > 0$. Also, $|K|$ has only one component since $H_0(K,\mathbb{Q}) \cong \mathbb{Q}$. But $H_0(K,\mathbb{Q})$ is generated by the homology class of *any* vertex of K, and therefore $f_{0*}^h:\mathbb{Q} \to \mathbb{Q}$ is the identity linear transformation. This shows that $\Lambda_f = 1$, so f has a fixed point.

As a direct corollary, we have a second proof of the Brouwer fixed-point theorem and, even better, we see that any contractible compact triangulable space has the fixed-point property. Now remember that the integral homology groups of the projective plane P^2 are $H_0(P^2) \cong \mathbb{Z}$, $H_1(P^2) \cong \mathbb{Z}_2$, and

$H_q(P^2) = 0$ for $q \geqslant 2$. Therefore the rational homology groups are the same as those of a point and we deduce that any map from the projective plane to itself must have a fixed point.

If X is a compact triangulable space, the Lefschetz number of the identity map 1_X is the Euler characteristic of X, by theorem (9.7). Since homotopic maps have the same Lefschetz number, the next result follows immediately:

(9.19) Theorem. *If the identity map of* X *is homotopic to a fixed-point free map then* $\chi(X) = 0$.

So the only closed surfaces which admit a fixed-point-free map that is homo-topic to the identity are the torus and Klein bottle. This proves our claim, made in Section 9.1, that the only hairy orientable surface which can be combed smooth is the torus, because by moving each point slightly along the hair growing out from it, we can produce a map without fixed points which is homotopic to the identity.

Finally, suppose we have a map $f: S^n \to S^n$. The only nonzero rational homology groups of S^n are $\mathbb{Q}$ in dimensions 0 and n, and in dimension n the homomorphism induced by f is just multiplication by the degree of f. We therefore have the following formula for the Lefschetz number of f.

(9.20) Theorem. $\Lambda_f = 1 + (-1)^n \deg f$.

From this formula we see that a map from S^n to S^n which does not have degree ± 1 must have a fixed point. Motivated by theorem (9.1), we shall call a homeomorphism $h: S^n \to S^n$ *orientation preserving* if the degree of h is $+1$, and *orientation reversing* if its degree is -1. If n is even (odd) then any orientation-preserving (reversing) homeomorphism of S^n has a fixed point, by the above formula.

Problems

25. If X is a compact triangulable space, and if $f: X \to X$ is null homotopic, show that f must have a fixed point.

26. Let G be a path-connected topological group. Show that left translation $L_g: G \to G$ by an element $g \in G$ is homotopic to the identity. (Notice you can join g to e by a path.)

27. Show that the Euler characteristic of a compact, connected, triangulable topological group is zero.

28. Show that the torus is the only closed surface which is a topological group.

29. Prove that an even-dimensional sphere cannot be a topological group. (In fact, S^1 and S^3 are the only spheres which are topological groups, though this is much harder to prove.)

30. Let K be a finite complex. If $f:|K| \rightarrow |K|$ is simplicial and has only isolated fixed points (in other words, the fixed points form a discrete set) show that Λ_f is the number of fixed points.

31. If K is a finite complex, and if $f:|K| \rightarrow |K|$ is simplicial, show that Λ_f is the Euler characteristic of the set of fixed points of f. (Remember the fixed points form a subcomplex of K^1.)

9.5 Dimension

We shall outline a method of defining the dimension of a compact Hausdorff space X. Let $\mathscr{F}$ be a *finite* open cover of X. Set $V = \mathscr{F}$, and agree that a collection $U_1,...,U_k$ of members of $\mathscr{F}$ belongs to S iff the intersection $U_1 \cap ... \cap U_k$ is nonempty. The hypotheses of the realization theorem (6.14) are easily checked, and realizing $\{V,S\}$ in a euclidean space gives a complex which we call the *nerve* of $\mathscr{F}$.

If K is a finite complex, and if $\mathscr{F}$ is the covering of $|K|$ by the open stars of the vertices of K, then the nerve of $\mathscr{F}$ is isomorphic to K by lemma (6.9). However, this example is not typical. Even if X is a triangulable space, the nerve of $\mathscr{F}$ may look nothing at all like X. Fig. 9.3 shows three open coverings of the circle. In the first case we obtain a 3-simplex plus its faces as nerve; and in the second case two vertices with a 1-simplex joining them. We do better with the third covering, whose nerve consists of four vertices and four 1-simplexes which fit together like the vertices and edges of a square. Here we have recaptured the topology of the circle.

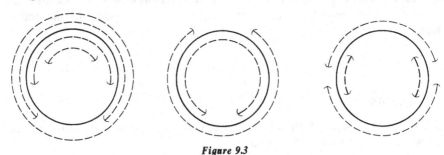

Figure 9.3

An open cover $\mathscr{F}'$ is a *refinement* of $\mathscr{F}$ if each member of $\mathscr{F}'$ is contained in some member of $\mathscr{F}$. So in the above example, the second cover refines the first, and the third refines both of the other two. The idea is that refining an open cover gives a better approximation to the original space X.

(9.21) Definition. *A compact Hausdorff space* X *has dimension* n *if every open cover of* X *has a refinement whose nerve has dimension at most* n, *and* n *is the smallest integer with this property.*

Since a homeomorphism from a space X to a space Y sends a finite open cover of X to one of Y, the dimension of a space is clearly a topological invariant of the space.

The definition of dimension given above is 'monotonic' in the following sense. If X and Y are compact Hausdorff spaces, and if Y is a subspace of X, then the dimension of Y is no more than that of X (Problem 34). Also, as we shall see below, our definition gives the correct answer for polyhedra.

Let K be a finite simplicial complex of dimension m and let $\mathscr{F}$ be a finite open cover of $|K|$. Using Lebesgue's lemma (3.11), we can find a barycentric subdivision K^r with the property that the open stars of the vertices of K^r form a refinement of $\mathscr{F}$. But the nerve of this covering of $|K|$ by open stars is isomorphic to K^r by lemma (6.9), and therefore has dimension m. This shows that the dimension of the space $|K|$ is no more than m.

We still have to check that the dimension of $|K|$ cannot be less than m. Now K contains an m-simplex Δ so, by monotonicity, it is enough to check that $|\Delta|$ has dimension m. Suppose the dimension of $|\Delta|$ is less than m, and let $\mathscr{F}$ be the open cover of $|\Delta|$ provided by the open stars of the vertices of Δ. Then $\mathscr{F}$ must have a refinement $\mathscr{F}'$ whose nerve is of dimension less than m. Choose a barycentric subdivision Δ^r with the property that the open stars of its vertices form a refinement $\mathscr{F}''$ of $\mathscr{F}'$. Write $N(\mathscr{F})$ for the nerve of $\mathscr{F}$. Since $\mathscr{F}'$ refines $\mathscr{F}$ we can define a simplicial map $s: |N(\mathscr{F}')| \to |N(\mathscr{F})|$ as follows. The vertices of $N(\mathscr{F}')$ are the open sets of $\mathscr{F}'$. If U is one of these open sets, choose V from $\mathscr{F}$ containing it and set $s(U) = V$. In exactly the same way, we have a simplicial map $t: |N(\mathscr{F}'')| \to |N(\mathscr{F}')|$. Now $N(\mathscr{F})$ is isomorphic to Δ and $N(\mathscr{F}'')$ to Δ^r, so the composition st is a simplicial map from $|\Delta^r|$ to $|\Delta|$. Also, the image of st has dimension less than m because st factors through $|N(\mathscr{F}')|$. So st is a map from $|\Delta|$ to $|\partial\Delta|$. Clearly $st|\,|\partial\Delta|$ is a null-homotopic map since it extends over $|\Delta|$. But, by its very construction, st is a simplicial approximation to the identity map from $|\Delta^r|$ to $|\Delta|$. Consequently, its restriction to $|\partial\Delta|$ cannot possibly be null homotopic, and we have the required contradiction.

We have proved the following result:

(9.22) Theorem. *If* K *is a finite simplicial complex of dimension* m, *its polyhedron* |K| *has dimension* m.

(9.23) Corollary. *The dimension of a finite simplicial complex is a topological invariant of its underlying polyhedron.*

Problems

32. Produce an open covering of the comb space whose nerve is a comb with only a finite number of teeth. What is the dimension of the comb space?

33. Where has homology theory been used in our proof of theorem (9.22)?

34. Show that our definition of dimension is 'monotonic' in the following sense. If X and Y are compact Hausdorff spaces, and if Y is a subspace of X, then the dimension of Y is no more than that of X. Where have you used the Hausdorff condition in your argument?

35. Define the dimension of a locally compact Hausdorff space to be that of its one-point compactification. Show that the dimension of the polyhedron of a (possibly infinite) complex is the dimension of the complex, and that this definition is monotonic.

36. What is the dimension of a discrete space?

37. Show that an n-manifold has dimension n.

10. Knots and Covering Spaces

But, as for everything else, so for a mathematical theory – beauty can be perceived but not explained.

A. CAYLEY

10.1 Examples of knots

We return to geometry in this chapter and consider various ways of embedding the circle as a subspace of $\mathbb{E}^3$. At first sight, the problem may seem rather narrow and special but, as we shall soon see, it is a meeting point for almost all the geometric and algebraic tools which we have developed so far.

A knot is a subspace of euclidean three-dimensional space which is homeomorphic to the circle. Fig. 10.1 illustrates four knots which happen to have special names; of course, in order to draw the knots we are forced to represent them by their projections in the plane of the paper. In addition, we mention the so-called trivial knot, or 'unknot', which consists of the unit circle in the (x, y) plane.

Trefoil

Figure of eight

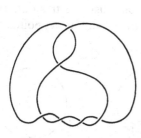

Stevedore's

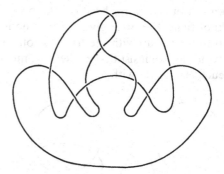

True lovers'

Figure 10.1

Suppose we make up each of the above knots using pieces of string (we urge the reader actually to do this). A little experiment quickly convinces us that we cannot convert any of these four into the trivial knot, nor indeed any one into another, simply by wobbling the string around. In order to do so we would have to let the string cross itself or (god forbid!) cut the string, make up our knot in a different way, and then tie the string up again. In some sense, which we need to make mathematically precise, these knots are all different.

The easiest way of saying when two knots are the same is to ask for a homeomorphism of 3-space which simply throws one knot onto the other, and this is the attitude we shall adopt here.

(10.1) Definition. *Two knots* k_1, k_2 *are equivalent if there is a homeomorphism* h *of* $\mathbb{E}^3$ *such that* $h(k_1) = k_2$.

We may be slightly disappointed that our definition says nothing about actually 'sliding' k_1 about in space until it lands up on top of k_2. In fact, the two ideas are not the same. Reflection in a plane is a perfectly good homeomorphism of $\mathbb{E}^3$ and transforms a knot to its mirror image. However, try as we may, we find that we cannot deform the trefoil knot into its mirror image (Fig. 10.2) without untying it.

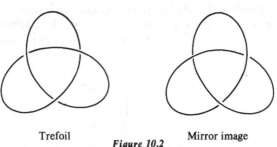

Trefoil   **Figure 10.2**   Mirror image

If we are looking for a definition of equivalence which involves sliding one knot around until it becomes the other, we have to be rather wary. Pulling a knot tight (Fig. 10.3) gives a continuous one-parameter family of knots which always ends up with the trivial knot, so any definition must rule this out. To avoid this we insist that as the knot moves it carries the neighbouring points of euclidean space with it.

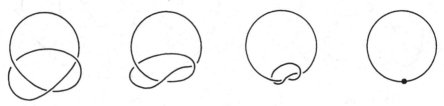

Figure 10.3

We say that a homeomorphism h of $\mathbb{E}^3$ is *isotopic to the identity* if there is a homotopy $H:\mathbb{E}^3 \times I \to \mathbb{E}^3$ such that each $h_t:\mathbb{E}^3 \to \mathbb{E}^3$ is a homeomorphism, h_0 is the identity, and $h_1 = h$. If we have a homeomorphism h which is isotopic to the identity and for which $h(k_1) = k_2$, then the knots $h_t(k_1)$ provide a continuous family which move gradually from k_1 to k_2 as t increases from 0 to 1.

If $h:\mathbb{E}^3 \to \mathbb{E}^3$ is a homeomorphism, we know that it extends in a unique way to a homeomorphism $\hat{h}:S^3 \to S^3$, because S^3 is the one-point compactification of $\mathbb{E}^3$. We say that h is orientation preserving or orientation reversing, according as $\hat{h}$ preserves or reverses the orientation of S^3. Now a homeomorphism which is isotopic to the identity must be orientation preserving. For we can extend each h_t to $\hat{h}_t:S^3 \to S^3$, and all we have to do is remember that homotopic maps have the same degree. On the other hand, reflection in a plane is orientation reversing and so cannot be isotopic to the identity. It is in fact true that any orientation-preserving homeomorphism of $\mathbb{E}^3$ is isotopic to the identity, though we shall not give a proof here.

A knot is *polygonal* if it is made up of a finite number of line segments. We shall work only with knots which are equivalent to polygonal knots, the so-called *tame* knots. An example of a wild knot (obtained by tying an infinite number of knots one after the other) is shown in Fig. 10.4, but the study of such knots is outside the scope of our work here.

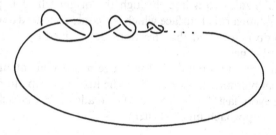

Figure 10.4

In order to picture knots and work with them effectively, we need to be able to project them into the plane in a nice way, the meaning of 'nice' being as shown in Fig. 10.1. The projection only crosses itself at a finite number of points, at most two pieces of the knot meet at such crossings, and they do so at 'right angles'. Our first result says that a polygonal knot always has a nice projection.

Let k be a polygonal knot. Given a direction, specified by a line in space, we can project k into the plane through the origin which is perpendicular to this direction. We call the projection *nice* if no more than two points of k map to each point of the image of k in the plane, the number of pairs of points of k identified by the projection is finite, and no such pair contains a vertex of k. A nice projection for a polygonal version of the figure-of-eight knot is shown in Fig. 10.5.

(10.2) Theorem. *Every polygonal knot has a nice projection.*

215

Proof. Certain directions have to be avoided. Firstly, those specified by prolonging the edges of k to give lines in $\mathbb{E}^3$; secondly, those determined by lines which join a vertex of k to an edge of k; and finally, those specified by lines which meet three edges of k. Now the set of all lines joining a given vertex

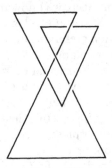

Figure 10.5

of k to a given edge determines a plane, and those lines which meet three skew edges of k form a ruled surface, called a regulus. By translating each of the generators of a regulus to a line through the origin which is parallel to the generator, we obtain a ruled surface which is a cone. So to find a nice projection, all we have to do is to avoid the directions determined by a finite number of lines, planes and cones.

We have not 'dotted all the i's' in the above proof. Our intention is to relax a little in this chapter and allow ourselves the luxury of explaining ideas rather than including every last detail of proof. The reader may well ask why we have not adopted this approach much earlier!

10.2   The knot group

If k_1, k_2 are equivalent knots we have a homeomorphism $h: \mathbb{E}^3 \to \mathbb{E}^3$ such that $h(k_1) = k_2$. Restricting h to $\mathbb{E}^3 - k_1$ gives a homeomorphism of $\mathbb{E}^3 - k_1$ with $\mathbb{E}^3 - k_2$, in other words, equivalent knots have homeomorphic complements. So it seems sensible to have a look at the fundamental group of the complement of a knot and see if we can use it to distinguish between various knots. Given a knot k, the fundamental group $\pi_1(\mathbb{E}^3 - k)$ is called the *knot group* of k. Our first job is to obtain some sort of reasonable presentation for a knot group in terms of generators and relations.

Take a copy of the knot in question in the upper half of 3-space and assume that projection into the plane $z = 0$ is nice. Break up the knot into 'overpasses' and 'underpasses', relative to this projection, which alternate as we go round the knot. Exactly how to do this is illustrated in Fig. 10.6 for the trefoil and the square knot, the overpasses are the heavier lines. Note that although we need to work with polygonal representations for our knots, we shall usually draw knots as smooth curves.

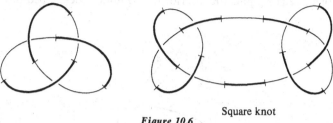

Square knot

Figure 10.6

Replace each underpass by the curve obtained on dropping perpendiculars from the endpoints of the underpass to the plane $z = 0$, then joining the free ends of these perpendiculars by the projection of the underpass. In this way, we obtain a new knot (Fig. 10.7) which is clearly equivalent to the original and

Figure 10.7

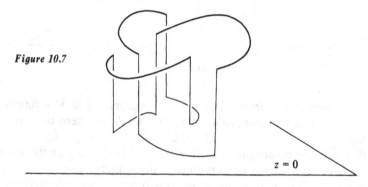

$z = 0$

which we shall denote by k. The idea is to calculate the knot group of k by building up $\mathbb{E}^3 - k$ out of several pieces, each of which has a fundamental group that we can recognize, and applying van Kampen's theorem (6.13) at each stage.

We first calculate $\pi_1(\mathbb{E}^3_+ - k)$, where $\mathbb{E}^3_+$ is the closed half-space defined by the inequality $z \geq 0$. Give a sense of direction to k and choose a base point p high in the air above k. For each overpass introduce a loop, which is based at p, and which winds once round the overpass in the sense of a right-hand screw relative to the direction of k, as shown in Fig. 10.8. Call these loops $\alpha_1, \ldots, \alpha_n$ and write x_i for the element of $\pi_1(\mathbb{E}^3_+ - k)$ determined by α_i.

(10.3) Lemma. $\pi_1(\mathbb{E}^3_+ - k, p)$ *is the free group generated by* $x_1, \ldots, x_n$.

Proof. Let $\hat{k}$ denote the overpasses of k plus the vertical line segments which join their end points to the plane $z = 0$. Then clearly $\mathbb{E}^3_+ - k$ and $\mathbb{E}^3_+ - \hat{k}$ have the same fundamental group. For each overpass we build a vertical wall up from the plane $z = 0$ to fit exactly underneath it, and we thicken this wall

217

slightly in $\mathbb{E}^3_+$ to give a three-dimensional ball (Fig. 10.9). We do this in such a way that the resulting balls $B_1,...,B_n$ are all disjoint. Suppose we now remove the interior of each B_i, plus the interior of the horseshoe-shaped disc in which

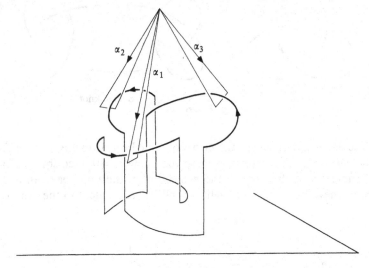

Figure 10.8

it meets the plane $z = 0$, from $\mathbb{E}^3_+$. Then the resulting space X is simply connected; actually it is homeomorphic to $\mathbb{E}^3_+$ but we do not need this much. We shall build up $\mathbb{E}^3_+ - \hat{k}$ as the union $X \cup (B_1 - \hat{k}) \cup ... \cup (B_n - \hat{k})$.

Any $B_i - \hat{k}$ is homeomorphic (Fig. 10.9) to a solid cylinder with its centre line removed. This deformation-retracts onto a disc minus its centre point, and therefore has fundamental group $\mathbb{Z}$ generated by a loop which links once

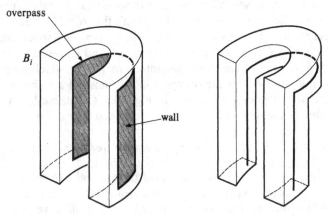

Figure 10.9

round k. Also, the intersection of $B_i - \hat{k}$ with X is homeomorphic to a disc and is therefore simply connected.

Suppose we know the fundamental group of $X \cup (B_1 - \hat{k}) \cup \ldots \cup (B_i - \hat{k})$ is the free group generated by $x_1,\ldots,x_i$. When we add in $B_{i+1} - \hat{k}$, van Kampen's theorem tells us we need an extra generator, which we can clearly take to be x_{i+1}. So we have an inductive proof of the lemma.

We end lemma (10.3) with a short dialogue:

Fussy Algebraist. You don't seem to care much about base points anymore.
Optimistic Geometer. They usually take care of themselves in this type of argument. Anyway I like drawing pictures, not worrying about base points.
F.A. To apply van Kampen's theorem there really should be a common base point in $[X \cup (B_1 - \hat{k}) \cup \ldots \cup (B_i - \hat{k})] \cap (B_{i+1} - \hat{k})$.
O.G. Here's an easy way out. For each i, join the base point p to some point on the top of B_i by a straight line and add this line to B_i. Now you really can base all the loops involved at p.
F.A. Even worse, you've only given a careful proof of van Kampen's theorem for finite simplicial complexes.
O.G. Convert $\mathbb{E}^3$ into S^3 by adding an extra point at ∞, and thicken k so as to give a tube T which is just a knotted solid torus in S^3. If we now replace $\mathbb{E}^3$ by S^3, $\mathbb{E}^3_+$ by the upper hemisphere, and if we remove the interior of the tube T whenever we should remove k, then all the spaces involved can be triangulated as finite simplicial complexes. But in terms of the fundamental group we have not changed anything because the extra point at ∞ is irrelevant, and because $T - k$ deformation retracts onto the boundary of T.

We still have all of $\mathbb{E}^3_- - k$ to add in. Suppose we look at the underpass of our knot which lies between the ith and $(i+1)$th overpass, and assume the kth overpass goes over it as in Fig. 10.10. Move the loops α_i, α_{i+1} close to the crossing, and take two loops α_k, $\bar{\alpha}_k$ to represent x_k, one on each side of the underpass as shown. Now thicken up the projection of the underpass in $\mathbb{E}^3_-$ to give a three-dimensional ball D_i, and consider the effect of adding $D_i - k$ to $\mathbb{E}^3_+ - k$. So that we can base all loops at p, we add to D_i a line which runs from p to q then vertically down to a point r on the top of D_i. Now $D_i - k$ is clearly simply connected, and $(D_i - k) \cap (\mathbb{E}^3_+ - k)$ consists of a disc with a polygonal arc removed from its interior. This latter has the same homotopy type as a disc with an interior point removed, and therefore has infinite cyclic fundamental group. If β_i is a loop which is based at p and which winds once round the projection of the underpass clockwise in the plane $z = 0$, then β_i represents a generator of this group which we shall denote by y_i.

According to van Kampen's theorem, if we want the fundamental group of $(\mathbb{E}^3_+ - k) \cup (D_i - k)$ we must add the relation $j_*(y_i) = e$ to $\pi_1(\mathbb{E}^3_+ - k)$, where j is the inclusion map of $(\mathbb{E}^3_+ - k) \cap (D_i - k)$ in $\mathbb{E}^3_+ - k$. But $j_*(y_i)$ is represented by the loop β_i *thought of as a loop in* $\mathbb{E}^3_+ - k$. By simply sliding β_i vertically upwards, we obtain a loop homotopic to the product loop $\alpha_i \alpha_k \alpha_{i+1}^{-1} \bar{\alpha}_k^{-1}$ (Fig.

10.10). So adding $D_i - k$ has the effect of imposing the relation $x_i x_k x_{i+1}^{-1} x_k^{-1} = e$ on $x_1, \ldots, x_m$, or equivalently

$$x_i x_k = x_k x_{i+1}$$

Note that if we reverse the direction of the kth overpass, then the relation changes to $x_k x_i = x_{i+1} x_k$.

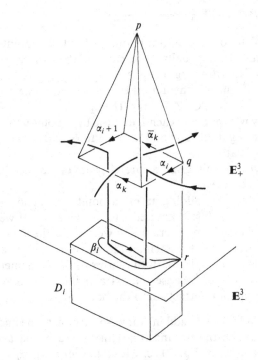

Figure 10.10

The other possibility is that our underpass is one which has been included simply to keep two overpasses apart. In this case, it should be clear that β_i is homotopic to $\alpha_i \alpha_{i+1}^{-1}$. So the extra relation to be added this time is $x_i = x_{i+1}$. Whichever relation we have, we denote it by the symbol r_i.

We have n underpasses altogether. The first $n-1$ give us relations $r_1, \ldots, r_{n-1}$ and tell us that the fundamental group of

$$Y = (\mathbb{E}^3_+ - k) \cup (D_1 - k) \cup \ldots \cup (D_{n-1} - k)$$

is $\{x_1, \ldots, x_n \mid r_1, \ldots, r_{n-1}\}$.

We claim that the relation corresponding to the final underpass is a consequence of the first $n-1$ and adds nothing new. For let Z denote the closure of $\mathbb{E}^3 - Y$. To complete our construction of $\mathbb{E}^3 - k$, all we have to do is to add $Z - k$ to Y. But $Z - k$ is simply connected, and $Y \cap (Z - k)$ has infinite cyclic fundamental group, generated by a loop which winds once round the projection of the final underpass. Now we are at liberty to choose this loop to be a large circle in the plane $z = 0$ which has the projection of our knot inside it. Simply sliding such a circle vertically upwards until it lies above k, then contracting it, shows that it represents the trivial element of $\pi_1(Y)$. A final application of van Kampen's theorem now gives our main result:

(10.4) Theorem. *The knot group of* k *is generated by the elements* $x_1,\ldots,x_n$ *subject to the relations* $r_1,\ldots,r_{n-1}$.

Here are some examples.

The trivial knot. We break up the circle into two semicircles, calling one an overpass. The above recipe then gives us one generator and no relations. *Therefore the knot group of the trivial knot is the infinite cyclic group.*

The trefoil. Take overpasses and underpasses as shown in Fig. 10.8. Then we have three generators x_1, x_2, x_3 subject to the relations $x_1 x_2 = x_3 x_1$, $x_2 x_3 = x_1 x_2$. Eliminating x_3 and writing $a = x_1, b = x_2$, this simplifies to give the group $G = \{a,b \mid aba = bab\}$.

Note that sending a to (12) and b to (23) defines a homomorphism from G to the symmetric group on three letters, since $(12)(23)(12) = (13) = (23)(12)(23)$. The homomorphism is onto because (12) and (23) generate the symmetric group S_3. This shows that G cannot be an abelian group; in particular, it cannot be $\mathbb{Z}$. We have therefore proved that the trefoil is not equivalent to the trivial knot. In other words *the trefoil really is knotted.*

The square knot. Take overpasses and underpasses as shown in Fig. 10.11, and label the underpasses 1 to 7. The letters a, b, c represent the generators of the knot group corresponding to three of the overpasses as shown, and using the relations given by underpasses 1, 2, 4, 5, we quickly express the other four generators in terms of these three. The relations corresponding to underpasses 3 and 6 are

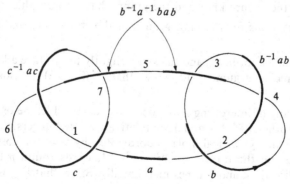

Figure 10.11

$$(b^{-1} a b)(b^{-1} a^{-1} b a b) = (b^{-1} a^{-1} b a b)b$$

which gives $aba = bab$, and

$$(c^{-1} a c)(b^{-1} a^{-1} b a b) = c(c^{-1} a c)$$

which reduces to $aca = cac$ when we replace bab by aba. The knot group of the square knot is therefore $\{a,b,c \mid aba = bab, \, aca = cac\}$.

Nonequivalent knots may have the same knot group. The left-hand half of the square knot looks like a trefoil, and the right-hand half like its mirror image. If we change the right-hand part to be a trefoil also, the resulting knot is called the granny, and is known to be a different knot. We ask the reader to compute the knot group of the granny and check it is isomorphic to that of the square knot.

Deciding whether or not two groups, given in terms of generators and relations, are isomorphic is in general impossible, and at best a painful task. For this reason, we would like a simpler invariant for distinguishing between knots, and one such will be introduced in Section 10.5. Note that abelianizing the knot group does not help. Looking at the form of the relations, we see that abelianizing simply sets all the generators equal to one another, giving the following result:

(10.5) Theorem. *Abelianizing a knot group always gives the infinite cyclic group.*

Problems

1. Find a presentation for the knot group of the figure-of-eight knot which has two generators. Show there is no homomorphism from this group onto the symmetric group S_3, and deduce that the figure of eight is not equivalent to the trefoil.

2. Check that the square knot and the granny have isomorphic knot groups.

3. Find presentations for the knot groups of the stevedore's and true lovers' knots.

4. Let k be a tame knot in $\mathbb{E}^3$, and thicken k slightly to produce a knotted tube T. Give a precise argument to show that $\pi_1(\mathbb{E}^3 - k)$ is isomorphic to $\pi_1(S^3 - \mathring{T})$.

5. A whole family of interesting knots occur as curves which lie on the surface of a standard torus in $\mathbb{E}^3$. If p and q are relatively prime integers, the *torus knot* $k_{p,q}$ is defined in cylindrical polar coordinates by $r = 2 + \cos(p\theta/q)$, $z = \sin(p\theta/q)$. It lies on the torus $(r - 2)^2 + z^2 = 1$, winds round p times in the longitudinal direction, and q times meridianally. Show that $k_{2,3}$ is the trefoil, and draw $k_{2,5}$.

6. Show that $k_{p,q}$ is equivalent to $k_{q,p}$ and to $k_{-p,q}$. Show also that $k_{p,q}$ is unknotted if $|p| = 1$ or $|q| = 1$.

7. Here are two ways of showing that the 3-sphere is the union of two solid tori:
(a) Prove that the set of points in $S^3 \subseteq E^4$ whose coordinates satisfy $x_1^2 + x_2^2 = x_3^2 + x_4^2$ is a torus, and that the inequalities $x_1^2 + x_2^2 \leqslant x_3^2 + x_4^2, x_1^2 + x_2^2 \geqslant x_3^2 + x_4^2$ both define solid tori.
(b) Think of S^3 as the join $S^1 * S^1$ of two circles. Show that the halfway section, which consists of points $tx + (1 - t)y$ for which $t = \frac{1}{2}$, is a torus, and that the inequalities $t \leqslant \frac{1}{2}, t \geqslant \frac{1}{2}$ both give solid tori.

8. Use Problem 7 and van Kampen's theorem to show that the knot group of $k_{p,q}$ has a presentation of the form $\{x,y \mid x^p = y^q\}$.

9. If G denotes the knot group of $k_{p,q}$, and H the subgroup generated by the element $x^p (= y^q)$, show that H is contained in the centre of G and that $G/H \cong Z_{|p|} * Z_{|q|}$.

10. Show that the free product of two nontrivial groups always has a trivial centre.

11. Assuming $|p| \neq 1, |q| \neq 1$, show that H is the centre of G. Now prove that if $1 < p < q, 1 < p' < q'$ then $k_{p,q}$ is equivalent to $k_{p',q'}$ iff $p = p'$ and $q = q'$.

12. Show that the set of equivalence classes of tame knots in E^3 is countable, but not finite.

10.3 Seifert surfaces

In this section we shall show how to span a tame knot k by an orientable surface. That is to say, we shall construct a compact, connected, orientable surface S in E^3 which has the knot k as boundary.

We illustrate the construction for the trefoil in Fig. 10.12. Orient the knot and choose a nice projection. Cut each crossing of the projection as shown to give a collection of disjoint oriented circles, called *Seifert circles*. Span each Seifert circle by a disc, keeping the discs disjoint, then replace the crossings by adding in a twisted strip at each crossing, as illustrated. The result is a compact connected surface S with boundary k. To see that S is orientable, notice that each Seifert circle has an orientation given from that of k. This determines an orientation for the disc which it spans, and the twisted strips are added in just such a way as to make these orientations all compatible with one another. S is called a *Seifert surface* for k.

Of course, we can span a given knot in many different ways; for example, take one Seifert surface and produce another by adding some handles to it well away from the knot. Now any Seifert surface S spanning k has boundary a single circle, so we can convert it into a *closed* surface by sewing a disc across this circle. By the genus of S we shall mean the genus of this orientable closed surface. One can read off the genus from a picture of the Seifert surface, because removing a disc from a closed orientable surface of genus g produces a space which deformation retracts onto the one-point union of $2g$ circles. The surface illustrated in Fig. 10.12 clearly deformation retracts onto the one-point union of two circles, so it has genus 1. Attaching a disc to it gives the torus.

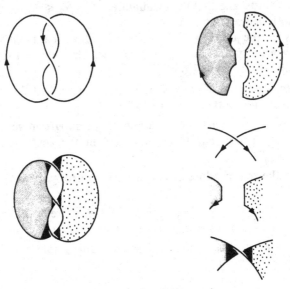

Figure 10.12

Call a surface S in $\mathbb{E}^3$ *tame* if there is a homeomorphism of $\mathbb{E}^3$ which throws S onto a finite simplicial complex, in other words onto a combinatorial surface (which may have a boundary). The smallest integer which occurs as the genus of a tame Seifert surface for k is called the *genus of k* and is written $g(k)$.

Our next result shows that having genus 0 is the same as being unknotted:

(10.6) Theorem. *A knot is equivalent to the trivial knot if and only if it can be spanned by a tame disc.*

Proof. Suppose k is equivalent to the unknot, and let h be a homeomorphism of $\mathbb{E}^3$ which throws k onto the boundary of the unit disc D in the (x,y) plane. Then $h^{-1}(D)$ is a tame disc spanning k.

Conversely, suppose k is polygonal and suppose we have a disc spanning k which is embedded polygonally in $\mathbb{E}^3$. In other words, the disc is chopped up into triangles, each of which lies linearly in $\mathbb{E}^3$. By a sequence of moves, each of which replaces one side of a triangle by the other two sides (or vice versa), we can change k to the boundary of a single triangle. But each such move can be realized by a homeomorphism of $\mathbb{E}^3$. Once we have thrown k onto the boundary of a triangle, it is a simple matter to find a homeomorphism which moves this triangle onto the unit disc in the plane. Therefore k is unknotted.

The genus of a knot is additive in the following sense. Suppose we have two *oriented* knots k and l which lie on opposite sides of a plane in $\mathbb{E}^3$, apart from a common arc in the plane on which they induce opposite orientations. Their *sum* $k + l$ is defined to be $k \cup l$ with all the points of this common arc, except the endpoints, removed (Fig. 10.13). Put in a less formal way, tie the knots one after the other in a piece of string, making sure their orientations agree. It is essential to work with oriented knots otherwise the definition may be ambiguous.

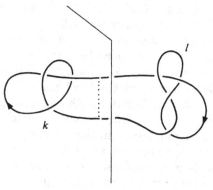

Figure 10.13

(10.7) Theorem. $g(k + l) = g(k) + g(l)$.

Sketch proof. First take copies of k and l which lie on opposite sides of a plane in $\mathbb{E}^3$, and span each of them by a (tame) Seifert surface of minimal genus in the appropriate half-space. Connect a little segment on k to one on l by a thin band, which is otherwise disjoint from the two spanning surfaces, and which twists if necessary so that the boundary of the resulting surface S is $k + l$ (Fig. 10.14). Clearly, the genus of S is the sum of the genus of the spanning surface for k with that for l. Therefore $g(k + l) \leqslant g(k) + g(l)$.

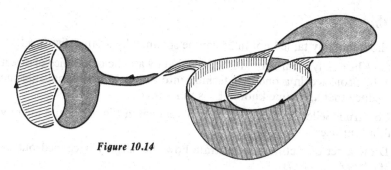

Figure 10.14

For the converse, begin with a tame Seifert surface S of minimal genus for $k + l$. We can always arrange that where S meets the plane P which separates k

from l it cuts through at 'right angles'. Therefore the intersection of S with P will be a collection of disjoint circles, plus an arc A whose endpoints are the points where $k + l$ pierces P. The idea is to do surgery on these circles, one by one, to produce a new minimal spanning surface which meets P only in A. Cutting along the arc A then gives Seifert surfaces for each of k and l whose genera add to the genus of S. Therefore $g(k) + g(l) \leqslant g(k + l)$.

The surgery goes as follows. Some of the circles of $S \cap P$ may be nested inside one another, and may even contain the arc A. Choose an innermost circle which does not contain A. Cut S along it and span the resulting circles by discs, one on each side of P. We can do this without running into other pieces of S because the circle in question does not contain any of the other circles, and does not contain A. The result must be a new surface which spans $k + l$ and meets P in one less circle, plus a closed surface which we ignore. (For if not, we have a single surface which spans $k + l$ and has smaller genus than before, by theorem (7.11), contradicting the fact that S has minimal genus.) We eliminate the intersection circles in this way. When we come to circles which contain the arc A, we start with the outermost one, cut along it, then cap off the two resulting circles by large discs which, in order to avoid S, go round behind k and l (just as if we blew up two balloons, pulled one over the part of S which is in the left-hand half-space, the other over the part on the right, and then attached their necks to the circles in question). Eventually, S meets P only in A, as required.

(10.8) Corollary. *If* $k + l$ *is equivalent to the trivial knot, then so is each of* k *and* l.

Proof. If $k + l$ is equivalent to the trivial knot, then $g(k) + g(l) = g(k + l) = 0$, giving $g(k) = g(l) = 0$. Therefore both k and l are unknotted by theorem (10.6).

This shows the impossibility of tying two knots in a row in a piece of string so that they cancel one another out.

Problems

13. Show that any tame knot in $\mathbb{E}^3$ can be spanned by a tame disc in $\mathbb{E}^4$.

14. Let k be a polygonal knot in $\mathbb{E}^4$. Show that by a judicious choice of direction, k can be projected in a one–one fashion into a three-dimensional subspace of $\mathbb{E}^4$. Deduce that any tame knot in $\mathbb{E}^4$ is unknotted.

15. Construct Seifert surfaces for the knots shown in Fig. 10.15 and identify the resulting surfaces.

16. Draw a set of pictures to illustrate how the surgery is carried out in the proof of theorem (10.7).

17. Show that neither the trefoil knot, nor the figure of eight, can be written as the sum of two nontrivial knots.

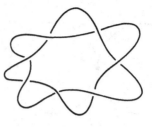

Figure 10.15

10.4   Covering spaces

The notion of a covering space was introduced rather briefly in Section 5.3. We plan to develop the idea a little further here, then have a look at a rather special covering space of the complement of a knot in the next section.

We recall the definition and one or two examples.

(10.9) Definition. *A map* $\pi: \tilde{X} \to X$ *is called a* covering map *and* $\tilde{X}$ *is said to be a* covering space *of* X *if the following condition holds. For each point* $x \in X$ *there is an open neighbourhood* V, *and a decomposition of* $\pi^{-1}(V)$ *as a family* $\{U_\alpha\}$ *of pairwise disjoint open subsets of* $\tilde{X}$, *in such a way that the restriction of* π *to each* U_α *is a homeomorphism from* U_α *to* V.

The exponential map from the real line to the unit circle in the complex plane is a covering map, as is the map from the 2-sphere to the projective plane obtained by identifying antipodal points. Both of these were considered in some detail in Chapter 5. Let n be a positive integer and consider the map from $\mathbb{C} - \{0\}$ to itself which raises each nonzero complex number to the nth power. This is a covering map (familiar from complex variable theory) which winds the punctured complex plane n times on itself.

Figure 10.16 shows a covering space of the one-point union of two circles. Reading from left to right, π winds the first circle of $\tilde{X}$ once round circle A, the second twice round circle B, the third twice round circle A, and so on. Note that exactly four points of $\tilde{X}$ map to each point of X. If x and V are as shown, then $\pi^{-1}(V)$ consists of the open sets U_i, $1 \leqslant i \leqslant 4$, each of which maps homeomorphically onto V under π. We leave the reader to choose appropriate neighbourhoods for points of B and for the point p where the two circles meet.

We shall assume that all our spaces are *path-connected* and *locally path-connected*. The latter condition (first introduced in Problem 43 of Chapter 3) simply means the topology on the space has a basis each of whose members is path-connected. For example, all polyhedra have this property.

Suppose then $\tilde{X}$ is a covering space of X with covering map $\pi: \tilde{X} \to X$.

227

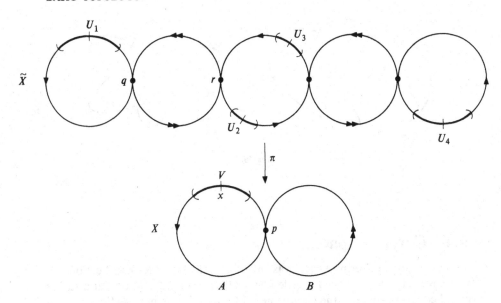

Figure 10.16

Choose base points $p \in X, q \in \tilde{X}$ with $\pi(q) = p$, and write $G = \pi_1(X, p)$, $H = \pi_1(\tilde{X}, q)$. We already have two basic results from our work in Chapter 5, lemmas (5.10) and (5.11):

(10.10) Path-lifting lemma. *If γ is a path in X which begins at p, there is a unique path $\tilde{\gamma}$ in $\tilde{X}$ which begins at q and satisfies $\pi \circ \tilde{\gamma} = \gamma$.*

(10.11) Homotopy-lifting lemma. *If $F: I \times I \to X$ is a map such that $F(0, t) = F(1, t) = p$ for $0 \leqslant t \leqslant 1$, there is a unique map $\tilde{F}: I \times I \to \tilde{X}$ which satisfies $\pi \circ \tilde{F} = F$ and $\tilde{F}(0, t) = q, 0 \leqslant t \leqslant 1$.*

Given $f: Y \to X$, a map $\tilde{f}: Y \to \tilde{X}$ with the property $\pi \circ \tilde{f} = f$ is usually called a *lift* of f. The fact that we can lift paths and homotopies into covering spaces has important consequences.

(10.12) Theorem. *The induced homomorphism $\pi_*: H \to G$ is one–one.*

Proof. Suppose $\tilde{\alpha}$ is a loop in $\tilde{X}$ based at q for which $\alpha = \pi \circ \tilde{\alpha}$ is null homotopic in X. Choose a specific homotopy F from the constant loop at p to α and apply lemma (10.11) to find $\tilde{F}: I \times I \to \tilde{X}$ satisfying $\pi \circ \tilde{F} = F$ and $\tilde{F}(0, t) = q$, $0 \leqslant t \leqslant 1$. Let P denote the union of the left- and right-hand edges and bottom of $I \times I$; then F maps all of P to p. But $\pi \circ \tilde{F} = F$, the set $\pi^{-1}(p)$ is a discrete set of points, and P is connected; therefore $\tilde{F}$ maps all of P to q. Also, the path in $\tilde{X}$ defined by $\tilde{F}(s, 1)$ is a lift of α which begins at q, and must therefore be $\tilde{\alpha}$ by

228

the uniqueness part of lemma (10.10). So $\tilde{F}$ is a homotopy from the constant loop at q to $\tilde{\alpha}$ as required.

(10.13) Theorem. *A loop α in X based at* p *lifts to a loop $\tilde{\alpha}$ in $\tilde{X}$ based at* q *if and only if $\langle \alpha \rangle \in \pi_*(H)$.*

Proof. One way is clear: if $\tilde{\alpha}$ is a loop then $\langle \alpha \rangle = \langle \pi \circ \tilde{\alpha} \rangle \in \pi_*(H)$. For the converse, suppose we have $\langle \alpha \rangle \in \pi_*(H)$; then we can find a loop β based at q in $\tilde{X}$ such that $\alpha \simeq \pi \circ \beta$. Choose a specific homotopy between these two loops and lift it into $\tilde{X}$ using lemma (10.11). An argument just like that of the proof of theorem (10.12) shows β and $\tilde{\alpha}$ must have the same endpoint, so $\tilde{\alpha}$ is a loop based at q.

Note that a loop in X may have one lift in $\tilde{X}$ which is a loop, and another which is a path with distinct endpoints. For example, the loop based at p in Fig. 10.16 represented by the circle A taken anticlockwise lifts to a loop based at q, yet the lift which begins at r is not a loop.

(10.14) Theorem. *For any point* x *of* X *the cardinality of the set $\pi^{-1}(x)$ is the index of $\pi_*(H)$ in G.*

Proof. First note that if $x,y \in X$ then $\pi^{-1}(x)$ and $\pi^{-1}(y)$ have the same cardinality. For let γ be a path in X which joins x to y. Given $\tilde{x} \in \pi^{-1}(x)$, lift γ to a path $\tilde{\gamma}$ in $\tilde{X}$ which begins at $\tilde{x}$. Then we have a function from $\pi^{-1}(x)$ to $\pi^{-1}(y)$ defined by $\tilde{x} \mapsto \tilde{\gamma}(1)$. This function must be one–one and onto, since we can produce an inverse for it using the path γ^{-1}.

Now consider $\pi^{-1}(p)$. Given a loop α in X based at p, lift it to a path $\tilde{\alpha}$ in $\tilde{X}$ which begins at q, and notice that $\tilde{\alpha}(1)$ is a point of $\pi^{-1}(p)$. If $\tilde{x} \in \pi^{-1}(p)$, projecting a path which joins q to $\tilde{x}$ into X gives a loop based at p, so every point of $\pi^{-1}(p)$ arises in this way. Now two loops α and β give the same point of $\pi^{-1}(p)$ if and only if $\alpha\beta^{-1}$ lifts to a loop based at q, and so by theorem (10.13), if and only if $\langle \alpha \rangle$ and $\langle \beta \rangle$ determine the same right coset of $\pi_*(H)$ in G. So we have a one–one onto correspondence from the right cosets of $\pi_*(H)$ in G to the set $\pi^{-1}(p)$.

If the inverse image of each point under π contains a finite number of points, say n, we say $\tilde{X}$ is an *n-sheeted* or *n-fold* covering space. For example, S^2 is a 2-sheeted covering of the projective plane; and the covering of $\mathbb{C} - \{0\}$ described earlier is n-sheeted. In the first case $H = \pi_1(S^2) = \{e\}$ and $G = \pi_1(P^2) \cong \mathbb{Z}_2$, therefore $\pi_*(H)$ has index 2 in G. In the second case, we have $H = G = \pi_1(\mathbb{C} - \{0\}) \cong \mathbb{Z}$ and $\pi_*(H) = n\mathbb{Z} \subseteq \mathbb{Z}$, so the index of $\pi_*(H)$ in G is indeed n.

(10.15) Theorem. *The groups $\pi_*(\pi_1(\tilde{X}, \tilde{x}))$, $\tilde{x} \in \pi^{-1}(p)$, form a conjugacy class of subgroups of* G.

Proof. The conjugacy class in question is that determined by $\pi_*(H)$. If

$\tilde{x} \in \pi^{-1}(p)$, join q to $\tilde{x}$ by a path $\tilde{\gamma}$ in $\tilde{X}$, write γ for the loop $\pi \circ \tilde{\gamma}$, and check that the following diagram commutes:

$$
\begin{array}{ccc}
H & \xrightarrow{\ \tilde{\gamma}_*\ } & \pi_1\,(\tilde{X},\tilde{x}) \\[2pt]
{\scriptstyle \pi_*}\Big\downarrow & & \Big\downarrow{\scriptstyle \pi_*} \\[2pt]
G & \xrightarrow{\ \gamma_*\ } & G
\end{array}
$$

So the inner automorphism of G given by γ_* throws $\pi_*(H)$ onto $\pi_*(\pi_1(\tilde{X}, \tilde{x}))$. For the converse, suppose $K = \langle\alpha\rangle^{-1}\pi_*(H)\langle\alpha\rangle$, where $\langle\alpha\rangle \in G$. Lift α to a path $\tilde{\alpha}$ in $\tilde{X}$ which begins at q, and set $\tilde{x} = \tilde{\alpha}(1)$. Then $\tilde{x} \in \pi^{-1}(p)$ and $K = \pi_*(\pi_1(\tilde{X},\tilde{x}))$.

So far, we have shown that a covering space of X picks out a conjugacy class of subgroups of the fundamental group of X. To make any further progress, we need a more general map-lifting result. Let Y be a space (path-connected and locally path-connected as always) with base point r, and let $f : Y \rightarrow X$ be a map which takes r to p.

(10.16) Map-lifting theorem. *There is a lift of* f *which takes* r *to* q *if and only if* $f_*(\pi_1(Y,r)) \subseteq \pi_*(H)$, *and this lift is unique.*

Proof. The necessity of the condition is clear since a lift $\tilde{f}$ gives a commutative diagram

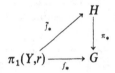

$$
\begin{array}{ccc}
 & & H \\[2pt]
 & {\scriptstyle \tilde{f}_*}\nearrow & \Big\downarrow{\scriptstyle \pi_*} \\[2pt]
\pi_1(Y,r) & \xrightarrow{\ f_*\ } & G
\end{array}
$$

Also, if we have a lift $\tilde{f}$ which satisfies $\tilde{f}(r) = q$, it must be unique. For suppose $\tilde{f}_1, \tilde{f}_2$ both lift f and send r to q. Given $y \in Y$, join r to y by a path γ. Then $\tilde{f}_1 \circ \gamma$ and $\tilde{f}_2 \circ \gamma$ are both lifts of $f \circ \gamma$ which begin at q, so they must agree, and in particular have the same final point. In other words, $\tilde{f}_1(y) = \tilde{f}_2(y)$.

Now suppose we have $f_*(\pi_1(Y,r)) \subseteq \pi_*(H)$, then we can construct $\tilde{f} : Y \rightarrow \tilde{X}$ as follows. Given $y \in Y$, join r to y by a path γ, lift the path $\alpha = f \circ \gamma$ to a path $\tilde{\alpha}$ in $\tilde{X}$ which begins at q, and set $\tilde{f}(y) = \tilde{\alpha}(1)$. The choice of γ does not matter for if γ' is a second path joining r to y, and if $\beta = f \circ \gamma'$ then $\alpha\beta^{-1}$ is a loop in X based at p. Also, $\langle\alpha\beta^{-1}\rangle$ lies in $f_*(\pi_1(Y,r))$ and therefore in $\pi_*(H)$. So by theorem (10.13), $\alpha\beta^{-1}$ lifts to a loop based at q in $\tilde{X}$. But for this to happen, $\tilde{\alpha}$ and $\tilde{\beta}$ must have the same final point.

We are left to check the continuity of $\tilde{f}$. Suppose $\tilde{f}(y) = \tilde{x}$ and $\pi(\tilde{x}) = x$, and let N be a neighbourhood of $\tilde{x}$ in $\tilde{X}$. Choose a neighbourhood V of x and a neighbourhood U of $\tilde{x}$ such that $\pi\,|\,U : U \rightarrow V$ is a homeomorphism. Then $f^{-1}\pi(N \cap U)$ is a neighbourhood of y in Y. Using the fact that Y is locally path-connected choose a *path-connected* neighbourhood W of y inside

$f^{-1}\pi(N \cap U)$. We claim that $\tilde{f}(W) \subseteq N$, and if we can prove this we are finished. Let $z \in W$, and join y to z in W by a path σ. To find $\tilde{f}(z)$, we lift the path $f \circ (\gamma\sigma) = (f \circ \gamma)(f \circ \sigma)$ to a path which begins at q in $\tilde{X}$, and take the endpoint of this path. Now $f \circ \sigma$ lies inside $\pi(N \cap U)$, and its lift has to start at the endpoint of the lift of $f \circ \gamma$, which is $\tilde{x}$. But $\pi \,|\, N \cap U$ is a homeomorphism, so the endpoint of this lift lies in $N \cap U$, and therefore in N as required.

We are now in a position to produce a hierarchical structure for the covering spaces of a given space. Let $\pi_1 : \tilde{X}_1 \to X$, $\pi_2 : \tilde{X}_2 \to X$ be covering maps. Choose base points $q_1 \in \tilde{X}_1, q_2 \in \tilde{X}_2$ so that $\pi_1(q_1) = \pi_2(q_2) = p$ and write $H_1 = \pi_1(\tilde{X}_1, q_1)$, $H_2 = \pi_1(\tilde{X}_2, q_2)$.

(10.17) Theorem. *If $\pi_{2*}(H_2) \subseteq \pi_{1*}(H_1)$ there is a covering map $\pi : \tilde{X}_2 \to \tilde{X}_1$ which sends q_2 to q_1 and satisfies $\pi_1 \circ \pi = \pi_2$.*

Proof. Simply apply theorem (10.16) to lift the map $\pi_2 : \tilde{X}_2 \to X$ to a map $\pi : \tilde{X}_2 \to \tilde{X}_1$ which sends q_2 to q_1, then check that π is a covering map.

Of course, if $\pi_{2*}(H_2)$ happens to equal $\pi_{1*}(H_1)$, then we can play this game in both directions and find basepoint-preserving covering maps $g : \tilde{X}_2 \to \tilde{X}_1$, $h : \tilde{X}_1 \to \tilde{X}_2$ which satisfy $\pi_1 \circ g = \pi_2$ and $\pi_2 \circ h = \pi_1$. Now $\pi_1 \circ g \circ h = \pi_1$, so $g \circ h$ and $1_{\tilde{X}_1}$ both lift $\pi_1 : \tilde{X}_1 \to X$ to a map from $\tilde{X}_1$ to $\tilde{X}_1$ which sends q_1 to q_1. By the uniqueness part of theorem (10.16), we must have $g \circ h = 1_{\tilde{X}_1}$. Similarly, $h \circ g = 1_{\tilde{X}_2}$, and we see that $h : \tilde{X}_1 \to \tilde{X}_2$ is a homeomorphism.

We shall call two covering spaces $\tilde{X}_1, \tilde{X}_2$ *equivalent* if we can find a homeomorphism $h : \tilde{X}_1 \to \tilde{X}_2$ such that $\pi_2 \circ h = \pi_1$. Combining the above discussion with theorem (10.15), we see that *two covering spaces of X are equivalent if and only if they determine the same conjugacy class of subgroups of the fundamental group of X.*

Now let $\pi : \tilde{X} \to X$ be a covering map, and define a *covering transformation* of $\tilde{X}$ to be a homeomorphism $h : \tilde{X} \to \tilde{X}$ which satisfies $\pi \circ h = \pi$. In the case of the covering of the projective plane by S^2, there are precisely two covering transformations, namely the identity map of S^2 and the antipodal map. For the covering $\pi : \mathbb{E}^1 \to S^1$ defined by $\pi(x) = e^{2\pi i x}$, a typical covering transformation is a translation of the real line by an integer. The set of all covering transformations of $\tilde{X}$ forms a group K under composition of homeomorphisms, and K acts freely on $\tilde{X}$. (For if h is a covering transformation of $\tilde{X}$, and if $h(\tilde{x}) = \tilde{x}$, then both h and $1_{\tilde{X}}$ agree on the point $\tilde{x}$ and lift $\pi : \tilde{X} \to X$. So they must be equal.)

(10.18) Theorem. *If $\pi_*(H)$ is a normal subgroup of G then X is homeomorphic to the orbit space $\tilde{X}/K$, and K is isomorphic to the factor group $G/\pi_*(H)$.*

Proof. The covering map $\pi : \tilde{X} \to X$ and the natural projection of $\tilde{X}$ onto $\tilde{X}/K$ are both identification maps, so we must check that the orbits of K are precisely the inverse images of points of X under π. Given $x \in X$, we know that each member of K permutes the points of $\pi^{-1}(x)$ because it is a covering transformation. Also, if $\tilde{x}, \tilde{y} \in \pi^{-1}(x)$, then $\pi_*(\pi_1(\tilde{X}, \tilde{x})) = \pi_*(H) = \pi_*(\pi_1(\tilde{X},\tilde{y}))$ by theorem (10.15) and the fact that $\pi_*(H)$ is normal in G. Therefore we can find a covering transformation which sends $\tilde{x}$ to $\tilde{y}$.

We are left to show K is isomorphic to $G/\pi_*(H)$. Given a loop α based at p in X, lift it to a path $\tilde{\alpha}$ which begins at q in $\tilde{X}$, and let k_α denote the unique covering transformation which sends q to $\tilde{\alpha}(1)$. Clearly, we can produce every element of K in this way, and by theorem (10.13) two loops α and β give the same element of K if and only if $\langle \alpha\beta^{-1} \rangle \in \pi_*(H)$. Therefore the correspondence $\alpha \mapsto k_\alpha$, induces a one–one onto function from $G/\pi_*(H)$ to K. To see this is a homomorphism, note that given two loops α, β based at p, the lift of $\alpha.\beta$ which begins at q is the path $\tilde{\alpha}.(k_\alpha \circ \tilde{\beta})$, and the endpoint of this is $k_\alpha(k_\beta(q))$. In other words, $\alpha.\beta$ corresponds to $k_\alpha \circ k_\beta$.

If $\pi_*(H)$ is a normal subgroup of G, we call $\tilde{X}$ a *regular* covering space. If a covering is not regular, there may not be enough covering transformations to go round, in the sense that we can have two points which map to the same point of X under π, and yet be unable to find a covering transformation which maps one to the other. The covering space of Fig. 10.16 is a good illustration of this situation. Here there is only one non-identity covering transformation; it acts as the antipodal map on the middle circle of $\tilde{X}$ and interchanges the two circles on the left with those on the right. In particular, no covering transformation maps q to r. Note that K is isomorphic to $\mathbb{Z}_2$, whereas the covering is 4-sheeted. We leave the reader to check that the orbit space $\tilde{X}/K$ consists of three circles joined together in a row.

Suppose $\tilde{X}$ is a simply connected covering space of X, then it is unique up to homeomorphism, since any two such must be equivalent, and it is a regular covering space of any other covering space of X by theorem (10.17). For these reasons, $\tilde{X}$ is called the *universal covering space* of X. Here are some examples. The universal covering space of the circle is the real line; that of projective n-space is S^n; that of the Klein bottle is the plane; and finally, that of the one-point union of two circles is the universal television aerial described in Chapter 6 (Fig. 6.21).

Suppose X has a universal covering space, and denote it by $\tilde{X}$. Then the covering transformations form a group isomorphic to the fundamental group of X. Given any subgroup H of $\pi_1(X)$, it acts on $\tilde{X}$ and the associated orbit space $\tilde{X}/H$ is a covering space of X whose fundamental group is isomorphic to H. So if X has a universal covering space, it has a covering space which corresponds to any subgroup of its fundamental group.

In order to ensure the existence of a universal covering space for a space X, we need to impose an extra condition on X. Call X *semi-locally simply connected* if each point of X has a neighbourhood U such that each loop in U is null homotopic in X. This is true of any polyhedron, but not for example of the Hawaiian earring (see Chapter 4, Problem 5).

(10.19) Existence theorem. *A space which is path-connected, locally path-connected, and semi-locally simply connected has a universal covering space.*

Proof. The details are rather long, but not difficult, so we give only the idea. Choose a base point p in X. The points of $\tilde{X}$ are equivalence classes of paths in X which begin at p, two such paths α, β being understood to be equivalent iff they have the same endpoint and $\alpha\beta^{-1}$ is homotopic to the constant loop at p.

To define $\pi:\tilde{X} \to X$, represent $\tilde{x} \in \tilde{X}$ by an appropriate path α in X, and let $\pi(\tilde{x})$ be the endpoint of α.

To construct a basis for the topology of $\tilde{X}$, begin with a path-connected open set V in X such that any loop in V is null homotopic in X, and a point $\tilde{x} \in \tilde{X}$ such that $\pi(\tilde{x}) \in V$. Represent $\tilde{x}$ by a path α in X which joins p to $\pi(\tilde{x})$, and define $V_{\tilde{x}}$ to be the subset of $\tilde{X}$ determined by paths in X of the form $\alpha.\beta$, where β lies in V. These sets $V_{\tilde{x}}$ are the basic open sets.

We leave the reader to check that $\pi:\tilde{X} \to X$ is a covering map, and that $\tilde{X}$ is path-connected and simply connected. No new ideas are involved, though we should point out that having chosen V so that all loops in V are null homotopic in X ensures that $\pi \mid V_{\tilde{x}}:V_{\tilde{x}} \to V$ is one–one.

Problems

18. If $\tilde{X}$ is a covering space of X, and $\tilde{Y}$ a covering space of Y, show that $\tilde{X} \times \tilde{Y}$ is a covering space of $X \times Y$.

19. Is the map $f:(0,3) \to S^1$ defined by $f(x) = e^{2\pi ix}$ a covering map?

20. Describe all the covering spaces of the torus, projective plane, Klein bottle, Möbius strip, and cylinder.

21. Find the group of covering transformations for each of the coverings of Problem 20.

22. The space shown in Fig. 6.19 is a covering space of the one-point union of two circles. What is the corresponding subgroup of $\mathbb{Z} * \mathbb{Z}$?

23. If X is a connected, locally path-connected, Hausdorff space, and if a finite group G of order n acts freely on X, show that X is an n-sheeted covering of X/G.

24. Assuming $p \geq 3$, find a fixed-point-free action of Z_{p-1} on $H(p)$ which has orbit space $H(2)$. Deduce that $H(p)$ is a $(p-1)$-fold covering space of $H(2)$. You may find Fig. 7.22 very helpful.

25. Formulate a similar result to that of Problem 24 for nonorientable surfaces.

26. Let $\pi:\tilde{G} \to G$ be a covering map, and suppose G is a topological group. Find a multiplication on $\tilde{G}$ which makes it into a topological group, and for which π is a homomorphism. Now show that the kernel of π is a discrete subgroup of $\tilde{G}$.

27. Examine the examples of group actions given in Section 4.4, and for each one deduce whether or not the associated projection is a covering map.

28. Prove that every closed nonorientable surface has a 2-sheeted orientable covering space.

10.5 The Alexander polynomial

The object of this final section is to produce a knot invariant in the form of a polynomial with integer coefficients, and to give a very simple algorithm for computing it. We shall explain the theory behind the polynomial first, the algorithm comes later!

Let k be a tame knot and, for convenience, think of k as a subset of the 3-sphere S^3. Thicken k slightly to form a knotted tube T, and let X denote S^3 with the interior of T removed. We refer to X as the complement of k: it has the advantage of being compact, and can be triangulated by a finite complex. Let G denote the knot group of k, in other words the fundamental group of X, and write G' for its commutator subgroup. By theorem (10.5), G/G' is the infinite cyclic group. If $\tilde{X}$ denotes the regular covering space of X which corresponds to G', we know that $\pi_1(\tilde{X})$ is isomorphic to G', and that the group of covering transformations of $\tilde{X}$ is infinite cyclic. $\tilde{X}$ is called the *infinite cyclic covering space* of X.

The existence of such a covering space $\tilde{X}$ follows from theorem (10.19), but to give ourselves a better feeling for it, and to convince ourselves that we can triangulate it as an infinite complex, we shall explain a simple method for constructing it. Very briefly, find a Seifert surface S for k (tame as always), and triangulate S^3 so that k, T, and S are all subcomplexes. Now cut X open along S. (This is not hard to visualize. If we have a triangulated surface, and a curve on the surface which is a subcomplex, we can imagine cutting the surface open along the curve. Each 1-simplex on the curve gives a pair of 1-simplexes when we cut, which have to be glued together again if we want to recapture the original surface. Our situation is just one dimension up from this.) When we cut X open along S, each triangle of S becomes a pair of triangles, and we label one set of triangles with the number 1, the other with the number 2, to remember which is which. (If you don't like cutting things open, here is an alternative. Begin with the disjoint union of all the 3-simplexes in X, and glue two together iff they have a triangle in common in X which does not lie in the surface S.) Denote the resulting simplicial complex by Y. Take a countable number of copies ... $Y_{-1}, Y_0, Y_1, Y_2, \ldots$ of Y, and glue them together as follows. Any triangle labelled 1 in Y_i should be glued to the corresponding triangle labelled 2 in Y_{i+1}. Write $\tilde{X}$ for the resulting space, and note that $\tilde{X}$ is triangulated as an infinite simplicial complex.

There is a natural map from each Y_i to X: simply glue up the simplexes which were separated when we cut along S, and these fit together to give a map $\pi:\tilde{X} \to X$, which is easily checked to be a covering map. The homeomorphism $h:\tilde{X} \to \tilde{X}$ which moves each point in Y_i to the corresponding point in Y_{i+1} generates an infinite cyclic group of homeomorphisms of $\tilde{X}$, and these are precisely the covering transformations. By theorem (10.18) we have $\pi_1(X)/\pi_*(\pi_1(\tilde{X})) \cong \mathbb{Z}$, and therefore $\pi_*(\pi_1(\tilde{X}))$ must contain the commutator subgroup G' of G. Now consider the natural epimorphism $\pi_1(X)/G' \to \pi_1(X)/\pi_*(\pi_1(\tilde{X}))$. It must be an isomorphism, since both groups are $\mathbb{Z}$. Since its kernel is $\pi_*(\pi_1(\tilde{X}))/G'$ we see that $\pi_*(\pi_1(\tilde{X})) = G'$. Therefore $\tilde{X}$ is the infinite cyclic covering space of X.

The next step is to take a look at the first homology group of $\tilde{X}$. Now $\tilde{X}$ is an infinite complex, so we should say what we mean by this. We start from the chain group whose elements are *finite* linear combinations $\lambda_1\sigma_1 + \ldots + \lambda_s\sigma_s$ of oriented 1-simplexes of $\tilde{X}$ with integer coefficients, agreeing that $(-\lambda)\sigma$ and $\lambda(-\sigma)$ always mean the same thing, and proceed exactly as in the case of a finite complex. Notice that we do not allow linear combinations of infinitely many 1-simplexes of $\tilde{X}$. The resulting homology group is an abelian group (though not finitely generated) which we denote by $H_1(\tilde{X})$. It is a topological invariant of $\tilde{X}$ because the proof of theorem (8.3) works as before and shows it to be the quotient of $\pi_1(\tilde{X})$ by its commutator subgroup. The covering transformation $h:\tilde{X} \to \tilde{X}$ induces an automorphism $h_*:H_1(\tilde{X}) \to H_1(\tilde{X})$, and it is h_* which will give us our polynomial.

At this point, we need a small dose of commutative algebra: a good elementary reference is Hartley and Hawkes [28]. Let Λ be a commutative ring with identity, and let A be an $m \times n$ matrix with entries from Λ. Write Λ^n for the free Λ module with basis $x_1,\ldots,x_n$, and Λ^m for the free Λ module with basis $y_1,\ldots,y_m$. Now let $f:\Lambda^n \to \Lambda^m$ be the Λ module homomorphism determined by A, in other words, that determined by the equations

$$f(x_i) = \sum_{j=1}^{m} a_{ji} y_j$$

and define M to be the quotient module $\Lambda^m/f(\Lambda^n)$. The matrix A is called a *presentation matrix* for the module M.

Two matrices A, B give isomorphic Λ modules under this construction iff we can convert A into B by a sequence of operations of the following type:

(a) interchange two rows or two columns;
(b) multiply a row, or a column, by a unit of Λ;
(c) add any multiple of one row to another row, or a multiple of one column to another column;
(d) add, or remove, a column of zeros;
(e) interchange A with $\begin{pmatrix} A & 0 \\ 0 & 1 \end{pmatrix}$, or vice versa.

Also, if A happens to be a square matrix, then its *determinant* is an isomorphism invariant of the module M. (This determinant is of course only determined up to multiplication by a unit of the ring Λ.)

Now back to the geometry. Let Λ be the ring $\mathbb{Z}[t,t^{-1}]$ of finite Laurent polynomials

235

$$p(t) = c_{-k} t^{-k} + \ldots + c_l t^l$$

with integer coefficients, and the usual rules for addition and multiplication of polynomials. Then h_* makes $H_1(\tilde{X})$ into a Λ module because we can define the product of a polynomial $p(t) \in \Lambda$ with a homology class $[z] \in H_1(\tilde{X})$ by

$$p(t)[z] = c_{-k} h_*^{-k}[z] + \ldots + c_l h_*^l[z]$$

We do not have a machine for computing homology groups, but we can interpret $H_1(\tilde{X})$ in terms of the knot group G of k, and we went to great lengths in Section 10.2 to produce a nice presentation for G. Remembering that $\tilde{X}$ has fundamental group G', and writing G'' for the commutator subgroup of G', we have $H_1(\tilde{X}) \cong G'/G''$. The monomorphism $\pi_* : \pi_1(\tilde{X}) \to \pi_1(X)$ induces an isomorphism $\pi_{**} : H_1(\tilde{X}) \to G'/G''$. If $u \in G'/G''$ corresponds to the homology class $[z]$ under this correspondence, then G'/G'' becomes a Λ module via $p(t)u = \pi_{**}(p(t)[z])$.

Let us examine what this formula means geometrically for the simple polynomial $p(t) = t$. Here we have

$$t\, u = \pi_{**}(t\,[z]) = \pi_{**} h_*[z]$$

Choose a base point $p \in X - S$, and let q be the corresponding point in $Y_0 \subseteq \tilde{X}$. Remembering always that $H_1(\tilde{X})$ is $\pi_1(\tilde{X},q)$ abelianized, represent $[z]$ by a loop α based at q in $\tilde{X}$. Then $h_*[z]$ is represented by the loop $\gamma(h \circ \alpha)\gamma^{-1}$, where γ is a path joining q to $h(q)$ in $\tilde{X}$. (The choice of γ is irrelevant.) This means that if $u = gG'' \in G'/G''$, and if x denotes the element of $G = \pi_1(X,p)$ determined by the loop $\pi \circ \gamma$, then

$$t\, u = x\, g\, x^{-1} \, G''$$

Notice that, by its construction, the homotopy class x goes to a generator of $\mathbb{Z}$ when we abelianize G.

The pieces of the jigsaw are now beginning to fit together. We recall the presentation $\{x_1, \ldots, x_n \mid r_1, \ldots, r_n\}$ of G developed in Section 10.2 from a projection of k. There is one generator for each overpass, and a relation for each crossing. A typical relation has the form $x_k x_i x_k^{-1} x_{i+1}^{-1}$, and the last relation can be omitted since it is a consequence of the others. When we abelianize G, all the x_i become equal and give a generator of $\mathbb{Z}$. So if we change to the new set of generators

$$x = x_n, \alpha_1 = x_1 x^{-1}, \ldots, \alpha_{n-1} = x_{n-1} x^{-1}$$

the elements $\alpha_1, \ldots, \alpha_{n-1}$ all lie in G', and it is not very hard to check that, together with all their conjugates under powers of x, they generate G'.

Write $u_i = \alpha_i G''$ and R_i for the relation among the u_i determined by r_i. The u_i generate G'/G'' as a module over Λ, and to find a presentation matrix all we have to do is to write out each R_i as a linear combination of the u_i, with coefficients from Λ, and then read off the coefficients and use them as the columns of our matrix. The determinant of the resulting $(n - 1) \times (n - 1)$ matrix is the famous *Alexander polynomial* of k.

For example, $x_k x_i x_k^{-1} x_{i+1}^{-1}$ when written in terms of the α_i becomes

$$\alpha_k x\alpha_i xx^{-1} \alpha_k^{-1} x^{-1} \alpha_{i+1}^{-1}$$

or equivalently

$$\alpha_k x\alpha_i x^{-1} x\alpha_k^{-1} x^{-1} \alpha_{i+1}^{-1}$$

Written additively, in terms of the u_i, this is

$$u_k + tu_i - tu_k - u_{i+1}$$

in other words

$$(1 - t)u_k + tu_i - u_{i+1}$$

We therefore have a column of the presentation matrix which has $1 - t$ in the kth place, t in the ith place, and -1 in the $(i + 1)$th place.

Suppose we do this for the trefoil. The group presentation is

$$\{x_1, x_2, x_3 \mid x_1 x_2 x_1^{-1} x_3^{-1}, x_2 x_3 x_2^{-1} x_1^{-1}\}$$

the first relation corresponds to $(1 - t)u_1 + tu_2$, the second to $(1 - t)u_2 - u_1$, and our prescription gives the matrix

$$\begin{pmatrix} 1 - t & -1 \\ t & 1 - t \end{pmatrix}$$

Therefore the Alexander polynomial of the trefoil knot is $t^2 - t + 1$. We comment again that the polynomial is only determined up to multiplication by a unit of Λ, in other words up to multiplication by $\pm t^k$.

It is not in fact necessary to work out the group presentation. Here is a purely formal algorithm for computing the Alexander polynomial from a nice projection of k. Orient the knot, and label the overpasses $x_1,...,x_n$. Construct an $n \times n$ matrix B which has a column for each crossing, the nonzero entries in the column corresponding to a crossing which looks like Fig. 10.17 being:

$1 - t$ in place k;
t in place i; and
-1 in place j.

Note that we only take into account the direction of x_k, not that of x_i and x_j. The determinant of the $(n - 1) \times (n - 1)$ matrix formed by removing the final row and final column from B is the Alexander polynomial of k. We need to

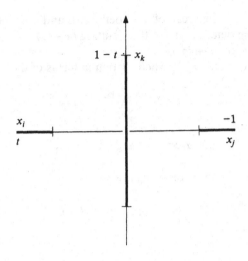

Figure 10.17

remove a row and a column (it does not matter which), because by using *all* the overpasses, and all the crossings, we have one too many generators and a redundant relation. In fact this matrix B is a presentation matrix for the direct sum of $H_1(\tilde{X})$ and a free Λ module of rank 1.

As an example, consider the stevedore's knot. Taking the crossings in the order indicated in Fig. 10.18 leads to the following matrix:

$$\begin{pmatrix} 1-t & -1 & 0 & 0 & -1 & 0 \\ 0 & t & 1-t & -1 & 0 & 0 \\ 0 & 0 & 0 & t & 1-t & t \\ 0 & 0 & -1 & 1-t & 0 & -1 \\ -1 & 1-t & t & 0 & 0 & 0 \\ t & 0 & 0 & 0 & t & 1-t \end{pmatrix}$$

and working out a 5×5 minor gives $2t^2 - 5t + 2$.

There is another, very elegant, description of the Alexander polynomial which we would like to mention, though the justification for it is too delicate to give here. Suppose we use rational coefficients, then $H_1(\tilde{X},\mathbb{Q})$ turns out to be a finite-dimensional vector space over $\mathbb{Q}$, and the Alexander polynomial is the characteristic polynomial of the linear transformation $h_*:H_1(\tilde{X},\mathbb{Q}) \to H_1(\tilde{X},\mathbb{Q})$.

238

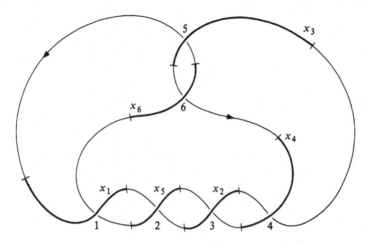

Figure 10.18

Problems

29. Show that the Alexander polynomial of the figure-of-eight knot is $t^2 - 3t + 1$. This shows the figure of eight to be knotted.

30. Work out the Alexander polynomials of the true lovers' knot and the two knots shown in Fig. 10.15. Check your answers against the tables given in Rolfsen [20].

31. Show that the Alexander polynomial of $k + l$ is that of k times that of l.

32. Suppose $\Delta(t)$ is the Alexander polynomial of a tame knot, and assume it has been normalized to have the form $a_0 + a_1 t + \ldots + a_k t^k$. Show that $\Delta(t) = t^k \Delta(1/t)$.

Appendix: Generators and relations

First courses in group theory traditionally take a student 'as far as' the classification theorem for finitely generated abelian groups, but invariably omit any discussion of free groups, or of the idea of presenting a group by means of generators and relations. Since these latter ideas are particularly important in topology (most especially for us in Chapters 6 and 10), we offer a quick survey here.

Perhaps the easiest idea to understand is that of a free set of generators for a given group. A subset X of a group G is called a *free set of generators* for G if every $g \in G - \{e\}$ can be expressed in a *unique* way as a product

$$g = x_1^{n_1} x_2^{n_2} \ldots x_k^{n_k} \tag{*}$$

of finite length, where the x_i lie in X, x_i is never equal to x_{i+1}, and each n_i is a nonzero integer. We call the set of generators free because by the uniqueness of (*) there can be no relations between its elements. If G has a free set of generators, then it is called a *free group*.

Given a nonempty set X, we can construct ourselves a group which has X as a free set of generators as follows. Define a *word* to be a finite product $x_1^{n_1} \ldots x_k^{n_k}$ in which each x_i belongs to X, and the n_i are all integers, and say that the word is *reduced* if x_i is never equal to x_{i+1} and all the n_i are nonzero. Given any word, we can make a reduced word out of it by collecting up powers when adjacent elements are equal, and omitting zeroth powers, continuing this process several times if necessary. An example is worth a page of explanation:

$$x_1^{-3} x_1^2 x_2^5 x_2^{-5} x_1^7 x_3^2 = x_1^{-1} x_2^0 x_1^7 x_3^2$$
$$= x_1^{-1} x_1^7 x_3^2$$
$$= x_1^6 x_3^2$$

which is now reduced. Reducing the word x_1^0 gives a word with no symbols which we refer to as the *empty word*. Now we can multiply words together simply by writing one after the other. If we do this with reduced words, the product may not be reduced, but it does simplify down to a well-defined reduced word which we call the *product* of the two given reduced words. The set of all reduced words forms a group under this multiplication (of course there is a lot of rather tedious checking to be done); the identity element is the empty word, and the inverse of the reduced word $x_1^{n_1}\ldots x_k^{n_k}$ is $x_k^{-n_k}\ldots x_1^{-n_1}$.

We shall call this group the *free group generated by* X, and denote it by $F(X)$. It should be clear that if two sets have the same cardinality (in other words, if there is a one–one onto correspondence between them) then the free groups generated by them are isomorphic. The free group with a single generator x is the infinite cyclic group, the only possible nonempty reduced words being the powers x^n.

Very often one says that a given group is determined by a set of generators and a set of relations. For example, we may say that the dihedral group with 10 elements is determined by two generators x, y subject to the relations $x^5 = e$, $y^2 = e$, $xy = yx^{-1}$. We have in mind an intuitive idea that all the elements of the group can be built as products of powers of x and y, and that the multiplication table of the group is completely specified by the given relations. We shall now make this precise using the notion of a free group.

Let G be a group, and X a subset which generates G. There is a natural homomorphism from the free group $F(X)$ onto G which sends a reduced word $x_1^{n_1}\ldots x_k^{n_k}$ onto the corresponding product of group elements in G (again we omit the details); it is onto because X generates G. If N denotes the kernel of this homomorphism, then $F(X)/N$ is isomorphic to G; so N determines G. Now let R be a collection of words in $F(X)$ with the property that N is the smallest normal subgroup containing them. These words, together with all their conjugates, generate N, and they determine exactly which words in $F(X)$ become the identity when we pass from $F(X)$ to G; that is to say, which products of elements of G are the identity in G. In this situation, we say that the pair X, R is a *presentation* for the group G. If X is a finite set, with elements $x_1,\ldots,x_m$, and R is a finite set of words, with elements $r_1,\ldots,r_n$, we say that G is finitely presented and write

$$G = \{x_1,\ldots,x_m \mid r_1,\ldots,r_n\}$$

Examples
1. $\mathbb{Z} = \{x \mid \varnothing\}$

2. $\mathbb{Z}_n = \{x \mid x^n\}$

3. The dihedral group with $2n$ elements is

$$D_{2n} = \{x,y \mid x^n, y^2, (xy)^2\}$$

4. $\mathbb{Z} \times \mathbb{Z} = \{x,y \mid x\,y\,x^{-1}\,y^{-1}\}$

We finish with a brief mention of free products. If G and H are groups we can form 'words' $x_1 x_2 \ldots x_m$, where each x_i lies in the disjoint union $G \cup H$. Call a word *reduced* this time if x_i and x_{i+1} never belong to the same group, and if x_i is never the identity of G or H. Throw in the empty word, multiply reduced words by juxtaposition, reducing the product as necessary, and the

242

result is a group called the *free product* $G * H$ of G and H. In this book, we only have occasion to take the free product of groups which are finitely presented, and we note that if

$$G = \{x_1,\ldots,x_m \mid r_1,\ldots,r_n\}, \qquad H = \{y_1,\ldots,y_k \mid s_1,\ldots,s_l\}$$

then

$$G * H = \{x_1,\ldots,x_m,y_1,\ldots,y_k \mid r_1,\ldots,r_n,s_1,\ldots,s_l\}$$

We note also that the free product $\mathbb{Z} * \mathbb{Z} * \ldots * \mathbb{Z}$ of n copies of the infinite cyclic group is just the free group on a set of size n.

The most important facts concerning free groups and free products are the following characterizations, which we give without proof:

(a) Let X be a subset of a group G. Then, X is a free set of generators for G iff given an arbitrary group K, plus a function from X to K, there is a *unique extension* of this function to a homomorphism from all of G to K.

(b) Let P be a group which contains both G and H as subgroups. Then P is isomorphic to the free product $G * H$, via an isomorphism which is the identity on both G and H, iff given an arbitrary group K, plus a homomorphism from each of G and H to K, there is a *unique extension* of these homomorphisms to a homomorphism from all of P to K.

Bibliography

Three classics
[1] Hilbert, D. and S. Cohn-Vossen, *Geometry and the Imagination*, Chelsea, New York, 1952.
[2] Lefschetz, S., *Introduction to Topology*, Princeton, 1949.
[3] Seifert, H. and W. Threlfall, *Lehrbuch der Topologie*, Teubner, Leipzig, 1934; Chelsea, New York, 1947.

Books at about the same level
[4] Agoston, M. K., *Algebraic Topology: A First Course*, Marcel Dekker, New York, 1976.
[5] Blackett, D. W., *Elementary Topology*, Academic Press, New York, 1967.
[6] Chinn, W. G. and N. E. Steenrod, *First Concepts of Topology*, Random House, New York, 1966.
[7] Crowell, R. H. and R. H. Fox, *Introduction to Knot Theory*, Ginn, Boston, 1963; Springer-Verlag, New York, 1977.
[8] Gramain, A., *Topologie des Surfaces*, Presses Universitaires de France, Paris, 1971.
[9] Massey, W. S., *Algebraic Topology: An Introduction*, Harcourt, Brace and World, 1967; Springer-Verlag, New York, 1977.
[10] Munkres, J. R., *Topology*, Prentice Hall, Englewood Cliffs, N.J., 1975.
[11] Singer, I. M. and J. A. Thorpe, *Lecture Notes on Elementary Topology and Geometry*, Scott Foresmann, Glenview, Ill., 1967; Springer-Verlag, New York, 1977.
[12] Wall, C. T. C., *A Geometric Introduction to Topology*, Addison Wesley, Reading, Mass., 1972.
[13] Wallace, A. H., *An Introduction to Algebraic Topology*, Pergamon, London, 1957.

More advanced texts
[14] Hilton, P. J. and S. Wylie, *Homology Theory*, Cambridge, 1960.
[15] Hirsch, M. W., *Differential Topology*, Springer-Verlag, New York, 1976.
[16] Hocking, J. G. and G. S. Young, *Topology*, Addison Wesley, Reading, Mass., 1961.
[17] Kelley, J. L., *General Topology*, Van Nostrand, Princeton, N.J., 1955; Springer-Verlag, New York, 1975.
[18] Maunder, C. R. F., *Algebraic Topology*, Van Nostrand Reinhold, London, 1970.
[19] Milnor, J., *Topology from the Differentiable Viewpoint*, University of Virginia Press, Charlottesville, 1966.
[20] Rolfsen, D., *Knots and Links*, Publish or Perish, Berkeley, 1976.
[21] Rourke, C. P. and B. J. Sanderson, *Piecewise Linear Topology*, Springer-Verlag, Berlin, 1972.
[22] Spanier, E. H., *Algebraic Topology*, McGraw-Hill, New York, 1966.

Papers

[23] Bing, R. H., 'The elusive fixed point property', *Amer. Math. Monthly*, **76**, 119–132, 1969.

[24] Doyle, P. H., 'Plane separation', *Proc. Camb. Phil. Soc.*, **64**, 291, 1968.

[25] Doyle, P. H. and D. A. Moran, 'A short proof that compact 2-manifolds can be triangulated', *Inventiones Math.*, **5**, 160–162, 1968.

[26] Tucker, A. W., 'Some topological properties of disc and sphere', *Proc. 1st Canad. Math. Congr.*, 285–309, 1945.

History

[27] Pont, J. C., *La Topologie Algebraique des Origines à Poincaré*, Presses Universitaires de France, Paris, 1974.

Algebra

[28] Hartley, B. and T. O. Hawkes, *Rings, Modules and Linear Algebra*, Chapman and Hall, London 1970.

[29] Lederman, W. *Introduction to Group Theory*, Longmans, London, 1976.

Comments

[1] is unbeatable for sheer enjoyment and has a chapter on elementary topology. Massey [9] is particularly good for surfaces, van Kampen's theorem, and covering spaces; his approach is different from ours, and his applications mainly directed towards proving results in group theory. Yet another way of classifying surfaces is provided by Gramain's elegant treatment in [8]. Algebraic topology at this level is nicely presented in [4] and [13], the first being particularly strong on applications and background history, and the second providing a contrasting approach with an account of singular homology.

Turning to more advanced material, for point-set topology [10], [16], and Kelley's classic [17] are very good indeed. In algebraic topology, the exact sequences of homology, cohomology, and duality are the next topics to look for. Of [14], [18], [22], Maunder is probably the easiest to break into. Finally, for topology with a more geometrical flavour we recommend [15], [20], [21], and especially [19].

Index

Undergraduate Texts in Mathematics

(continued from page ii)

Halmos: Finite-Dimensional Vector Spaces. Second edition.

Halmos: Naive Set Theory.

Hämmerlin/Hoffmann: Numerical Mathematics.
Readings in Mathematics.

Harris/Hirst/Mossinghoff: Combinatorics and Graph Theory.

Hartshorne: Geometry: Euclid and Beyond.

Hijab: Introduction to Calculus and Classical Analysis.

Hilton/Holton/Pedersen: Mathematical Reflections: In a Room with Many Mirrors.

Hilton/Holton/Pedersen: Mathematical Vistas: From a Room with Many Windows.

Iooss/Joseph: Elementary Stability and Bifurcation Theory. Second edition.

Isaac: The Pleasures of Probability.
Readings in Mathematics.

James: Topological and Uniform Spaces.

Jänich: Linear Algebra.

Jänich: Topology.

Jänich: Vector Analysis.

Kemeny/Snell: Finite Markov Chains.

Kinsey: Topology of Surfaces.

Klambauer: Aspects of Calculus.

Lang: A First Course in Calculus. Fifth edition.

Lang: Calculus of Several Variables. Third edition.

Lang: Introduction to Linear Algebra. Second edition.

Lang: Linear Algebra. Third edition.

Lang: Short Calculus: The Original Edition of "A First Course in Calculus."

Lang: Undergraduate Algebra. Second edition.

Lang: Undergraduate Analysis.

Lax/Burstein/Lax: Calculus with Applications and Computing. Volume 1.

LeCuyer: College Mathematics with APL.

Lidl/Pilz: Applied Abstract Algebra. Second edition.

Logan: Applied Partial Differential Equations.

Macki-Strauss: Introduction to Optimal Control Theory.

Malitz: Introduction to Mathematical Logic.

Marsden/Weinstein: Calculus I, II, III. Second edition.

Martin: Counting: The Art of Enumerative Combinatorics.

Martin: The Foundations of Geometry and the Non-Euclidean Plane.

Martin: Geometric Constructions.

Martin: Transformation Geometry: An Introduction to Symmetry.

Millman/Parker: Geometry: A Metric Approach with Models. Second edition.

Moschovakis: Notes on Set Theory.

Owen: A First Course in the Mathematical Foundations of Thermodynamics.

Palka: An Introduction to Complex Function Theory.

Pedrick: A First Course in Analysis.

Peressini/Sullivan/Uhl: The Mathematics of Nonlinear Programming.

Prenowitz/Jantosciak: Join Geometries.

Priestley: Calculus: A Liberal Art. Second edition.

Protter/Morrey: A First Course in Real Analysis. Second edition.

Protter/Morrey: Intermediate Calculus. Second edition.

Pugh: Real Mathematical Analysis

Roman: An Introduction to Coding and Information Theory.

Undergraduate Texts in Mathematics